KAPLAN
K12 LEARNING SERVICES

Kaplan
ADVANTAGE
ACT* Science

LEVEL HS

*ACT® is a registered trademark of ACT, Inc., which does not endorse this product.

Curriculum Development
Rebecca Machalow

Contributing Writers
Elizabeth Boskey, Nancy Costigliola, Maria Malzone, Adam Marks, Bhavana Nancherla, Allen Ruch

Design
Maurice Kessler

Production
Janice Loehrer, Dorothy Nocerino, Scott Rayow, Stephanie Rodriguez

Illustration
Thomas Kurzanski, Erika Quiroz, Theresa Seelye

Cover Design
Maurice Kessler

Production Manager
Michael Young

Managing Editor
Michelle K. Winberg

Editorial
Julie Freydlin, Maria Malzone

Project Coordinators
Sandra Ogle, Aaron Riccio

Director of Curriculum and Instruction
Deborah Lerman

Copyright © 2006 Kaplan, Inc.

All rights reserved. No part of this book may be reproduced or transmitted in any form or by any means, electronic or mechanical, including photocopying, recording, or by any information storage and retrieval system, without the written permission of the Publisher, except where permitted by law.

TABLE OF CONTENTS

Unit 1: *Getting to Know the ACT* 1
    LESSON A: Practice Test 1 3
    LESSON B: The Landscape of the Test 21
    LESSON C: Eliminating. 33

Unit 2: *Data Representation Passages* 47
    LESSON A: Find the Keywords 49
    LESSON B: Tables and Figures 63
    LESSON C: Using Patterns. 79

Unit 3: *Data Representation Passages Part 2* 93
    LESSON A: Understanding Experiments. 95
    LESSON B: Complex Figure Questions 109
    LESSON C: Equations in the Science ACT 123

Unit 4: *Conflicting Viewpoints* 139
    LESSON A: Questions About One Viewpoint . . . 141
    LESSON B: Analyzing the Passage. 155
    LESSON C: General Test-Taking Strategies 169

Practice Test 2 181

Unit 1
Getting to Know the ACT

lesson A
Practice Test 1

Practice Test 1

Welcome to *ACT Advantage*! This program will help you prepare for the ACT in Science. The first step in your climb to the summit of test success is a Practice Test.

This test serves two purposes. First, it will give you a feel for the types of questions on the ACT, and how much time you have to complete them. Second, it will give you and your teacher a sense of your strengths and weaknesses so that you can better prepare for Test Day. Your performance will not be graded or scored.

You will see seven passages on this Practice Test. Each passage will be followed by 5–7 questions. The ACT does not deduct points for incorrect answers, so if you are unsure about a question, just take your best guess.

Practice Test 1

When your teacher tells you, carefully tear out this page. Then begin working.

1. Ⓐ Ⓑ Ⓒ Ⓓ
2. Ⓕ Ⓖ Ⓗ Ⓙ
3. Ⓐ Ⓑ Ⓒ Ⓓ
4. Ⓕ Ⓖ Ⓗ Ⓙ
5. Ⓐ Ⓑ Ⓒ Ⓓ
6. Ⓕ Ⓖ Ⓗ Ⓙ
7. Ⓐ Ⓑ Ⓒ Ⓓ
8. Ⓕ Ⓖ Ⓗ Ⓙ
9. Ⓐ Ⓑ Ⓒ Ⓓ
10. Ⓕ Ⓖ Ⓗ Ⓙ
11. Ⓐ Ⓑ Ⓒ Ⓓ
12. Ⓕ Ⓖ Ⓗ Ⓙ
13. Ⓐ Ⓑ Ⓒ Ⓓ
14. Ⓕ Ⓖ Ⓗ Ⓙ

15. Ⓐ Ⓑ Ⓒ Ⓓ
16. Ⓕ Ⓖ Ⓗ Ⓙ
17. Ⓐ Ⓑ Ⓒ Ⓓ
18. Ⓕ Ⓖ Ⓗ Ⓙ
19. Ⓐ Ⓑ Ⓒ Ⓓ
20. Ⓕ Ⓖ Ⓗ Ⓙ
21. Ⓐ Ⓑ Ⓒ Ⓓ
22. Ⓕ Ⓖ Ⓗ Ⓙ
23. Ⓐ Ⓑ Ⓒ Ⓓ
24. Ⓕ Ⓖ Ⓗ Ⓙ
25. Ⓐ Ⓑ Ⓒ Ⓓ
26. Ⓕ Ⓖ Ⓗ Ⓙ
27. Ⓐ Ⓑ Ⓒ Ⓓ
28. Ⓕ Ⓖ Ⓗ Ⓙ

29. Ⓐ Ⓑ Ⓒ Ⓓ
30. Ⓕ Ⓖ Ⓗ Ⓙ
31. Ⓐ Ⓑ Ⓒ Ⓓ
32. Ⓕ Ⓖ Ⓗ Ⓙ
33. Ⓐ Ⓑ Ⓒ Ⓓ
34. Ⓕ Ⓖ Ⓗ Ⓙ
35. Ⓐ Ⓑ Ⓒ Ⓓ
36. Ⓕ Ⓖ Ⓗ Ⓙ
37. Ⓐ Ⓑ Ⓒ Ⓓ
38. Ⓕ Ⓖ Ⓗ Ⓙ
39. Ⓐ Ⓑ Ⓒ Ⓓ
40. Ⓕ Ⓖ Ⓗ Ⓙ

SCIENCE TEST
35 Minutes—40 Questions

DIRECTIONS: There are seven passages in this test. Each passage is followed by several questions. After reading a passage, choose the best answer to each question and fill in the corresponding oval on your answer sheet. You may refer to the passages as often as necessary.

You are NOT permitted to use a calculator on this test.

Passage I

A panel of engineers designed and built a pressurized structure to be used for shelter by geologists during extended research missions near the South Pole. The design consisted of four rooms, each with its own separate heating and air pressure control systems. During testing, the engineers found the daily average air temperature, in degrees Celsius (°C), and daily average air pressure, in millimeters of mercury (mm Hg), in each room. The data for the first five days of their study are given in Table 1 and Table 2.

Table 1

Day	Daily average temperature (°C)			
	Room 1	Room 2	Room 3	Room 4
1	19.64	19.08	18.67	18.03
2	20.15	19.20	18.46	18.11
3	20.81	19.19	18.62	18.32
4	21.06	19.51	19.08	18.91
5	21.14	19.48	18.60	18.58

Table 2

Day	Daily average air pressure (mm Hg)			
	Room 1	Room 2	Room 3	Room 4
1	748.2	759.6	760.0	745.2
2	752.6	762.0	758.7	750.3
3	753.3	760.2	756.5	760.4
4	760.1	750.8	755.4	756.8
5	758.7	757.9	754.0	759.5

1. In which room was the greatest difference between maximum and minimum daily average air pressure observed over the study's first five days?
   A. Room 1
   B. Room 2
   C. Room 3
   D. Room 4

2. Looking at the data in Table 1, do you find any overall trend in the variation in daily average temperature for all the rooms?
   F. It decreased from Day 1 to Day 5.
   G. It increased from Day 1 to Day 4.
   H. It decreased up to Day 3, then increased.
   J. There is no trend in the data.

GO ON TO THE NEXT PAGE.

3. Which of the following graphs best represents a plot of the daily average air temperature versus the daily average air pressure for Room 4?

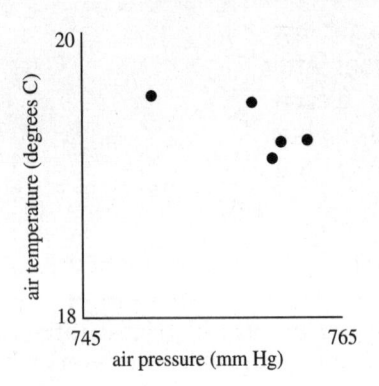

A.
B. 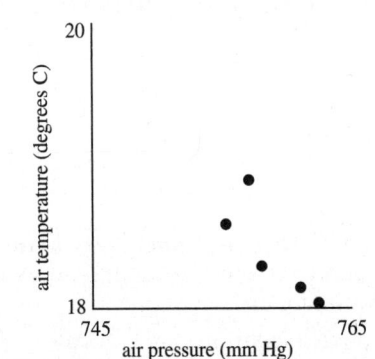
C. 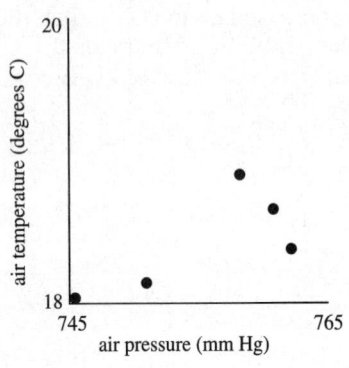
D. 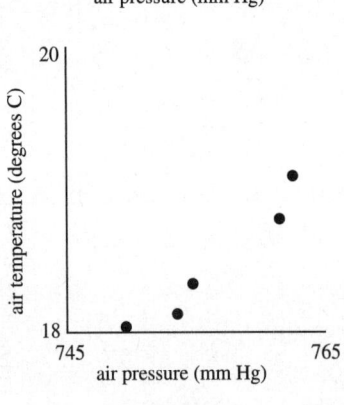

4. Which of the following most accurately describes the changes in the daily average air pressure in Room 3 during days 1–5?
   F. The daily average air pressure increased from days 1 to 4 and decreased from days 4 to 5.
   G. The daily average air pressure decreased from days 1 to 2, increased from days 2 to 4, and decreased again from days 4 to 5.
   H. The daily average air pressure increased only.
   J. The daily average air pressure decreased only.

5. Suppose the *heat absorption modulus* of a room is defined as the quantity of heat absorbed by the contents of the room, divided by the quantity of heat provided to the entire room. Based on the data, would one be justified in concluding that the heat absorption modulus of Room 1 was higher than the heat absorption modulus of any of the other rooms?
   A. Yes, because the quantity of heat provided to Room 1 was higher than the quantity of heat provided to any of the other rooms.
   B. Yes, because the quantity of heat not absorbed by the contents of Room 1 was higher than the quantity of heat not absorbed by the contents of any of the other rooms.
   C. No, because the quantity of heat absorbed by the contents of Room 1 was lower than the quantity of heat absorbed by the contents of one of the other rooms.
   D. No, because the information provided is insufficient to determine the heat absorption modulus.

GO ON TO THE NEXT PAGE.

Passage II

Humans can experience toxic symptoms when concentrations of mercury (Hg) in the blood exceed 200 parts per billion (ppb). Frequent consumption of foods high in Hg content contributes to high Hg levels in the blood. On average, higher Hg concentrations are observed in people whose diets consist of large amounts of certain types of seafood. A research group proposed that sea creatures that live in colder waters acquire greater amounts of Hg than those that reside in warmer waters. The researchers performed the following experiments to examine this hypothesis.

Experiment 1

Samples of several species of consumable sea life caught in the cold waters of the northern Atlantic Ocean were chemically prepared and analyzed using a cold vapor atomic fluorescence spectrometer (CVAFS), a device that indicates the relative concentrations of various elements and compounds found within a biological sample. Comparisons of the spectra taken from the seafood samples with those taken from samples of known Hg levels were made to determine the exact concentrations in ppb. Identical volumes of tissue from eight different specimens for each of four different species were tested, and the results are shown in Table 1, including the average concentrations found for each species.

Table 1				
Specimen	Hg concentrations in cold-water species (ppb):			
	Cod	Crab	Swordfish	Shark
1	160	138	871	859
2	123	143	905	820
3	139	152	902	839
4	116	177	881	851
5	130	133	875	818
6	134	148	880	836
7	151	147	910	847
8	109	168	894	825
Average	133	151	890	837

Experiment 2

Four species caught in the warmer waters of the Gulf of Mexico were examined using the procedure from Experiment 1. The results are shown in Table 2.

Table 2				
Specimen	Hg concentration in warm-water species (ppb):			
	Catfish	Crab	Swordfish	Shark
1	98	113	851	812
2	110	122	856	795
3	102	143	845	821
4	105	128	861	803
5	94	115	849	798
6	112	136	852	809
7	100	129	863	815
8	117	116	837	776
Average	105	125	852	804

6. Given that shark and swordfish are both large predatory animals, and catfish and crab are smaller non-predatory animals, do the results of Experiment 2 support the hypothesis that the tissues of larger predatory fish exhibit higher levels of Hg than do the tissues of smaller species?

F. Yes; the lowest concentration of Hg was found in swordfish.
G. Yes; both predators had higher Hg concentrations than those found in either catfish or crab.
H. No; the lowest concentration of Hg was in catfish.
J. No; both catfish and crab had higher Hg concentrations than those found in either predator.

GO ON TO THE NEXT PAGE.

7. A researcher, when using the CVAFS, was concerned that lead (Pb) in the tissue samples may be interfering with the detection of Hg. Which of the following procedures would best help the researcher explore this trouble?
   A. Flooding the sample with a large concentration of Pb before using the CVAFS
   B. Using the CVAFS to examine a non-biological sample
   C. Collecting tissue from additional species
   D. Testing samples with known concentrations of Hg and Pb

8. Based on the results of the experiments and the data in the table below, sharks caught in which of the following locations would most likely possess the largest concentrations of Hg in February?
   F. Northern Atlantic Ocean
   G. Northern Pacific Ocean
   H. Gulf of Mexico
   J. Tampa Bay

Location	Average water temperature (°F) for February
Northern Altantic Ocean	33
Gulf of Mexico	70
Northern Pacific Ocean	46
Tampa Bay	72

9. Looking at Experiment 2 as a stand-alone procedure, which of the following factors was intentionally varied?
   A. The volume of tissue tested
   B. The method by which the marine organisms were caught
   C. The species of marine organism tested
   D. The method of sample analysis

10. The governments of many nations require frequent testing of seafood to determine Hg concentration levels. According to the experiments, in order to determine the maximum concentration of Hg found in a collection of seafood specimens, from which of the following specimens would it be best to take sample tissue?
    F. A crab caught in cold water
    G. A swordfish caught in cold water
    H. A catfish caught in warm water
    J. A swordfish caught in warm water

11. How might the results of the experiments be affected if the scientist preparing the specimens for Experiment 2 introduced Hg-free contaminants into the samples, resulting in a larger volume of tested material?
    A. The actual concentrations of Hg would be lower for both cold-water and warm-water specimens.
    B. The actual concentrations of Hg would be higher for warm-water specimens than for cold-water specimens.
    C. The measured concentrations of Hg would be lower than the actual concentrations for warm-water specimens, but not for cold-water specimens.
    D. The measured concentrations of Hg would be higher than the actual concentrations for warm-water specimens, but not for cold-water specimens.

GO ON TO THE NEXT PAGE.

Passage III

A student performed three exercises with a battery and four different lightbulbs.

Exercise 1

The student connected the battery to a fixed outlet designed to accept any of the four bulbs. She then placed four identical light sensors at different distances from the outlet. Each sensor was designed so that a green indicator illuminated upon the sensor's detection of incident light, while a red indicator illuminated when no light was detected. The student darkened the room and recorded the color of each sensor while each bulb was lit. The results are shown in Table 1.

Table 1				
Sensor distance (cm)	Sensor indicator color			
	Bulb 1	Bulb 2	Bulb 3	Bulb 4
50	green	green	green	green
100	red	green	green	green
150	red	red	green	green
200	red	red	red	green

Exercise 2

The battery produced an *electromotive force* (a measure of voltage in an electrical circuit) of 12 Volts. The student was given a device called an ammeter, which is used to measure the *current* (the rate of flow of electric charge) passing through an electric circuit. She completed the circuit by connecting the battery, the ammeter, and one of the four lightbulbs. She measured the associated current in Amperes (A) and calculated the impedance (Z; the measure of resistance to electrical flow) in Ohms (Ω) for each lightbulb, using the following formula:

$$Z = \frac{\text{electromotive force}}{\text{current}}$$

The results are shown in Table 2.

Table 2		
Lightbulb	Current (A)	Z (Ω)
1	0.2	60
2	0.3	40
3	0.4	30
4	0.6	20

Exercise 3

The *power rating* (P) of each lightbulb was printed near its base. P gives the bulb's rate of energy consumption over time and is related to the *brightness* (B) of light at a given distance from the bulb. B is calculated in Watts per meters squared (W/m^2) from the following formula:

$$B = \frac{P}{4\pi r^2}$$

where r is the distance in meters (m) from the bulb, and P is measured in Watts (W).

The student calculated B for each bulb at a distance of 1 m. The results are shown in Table 3.

Table 3		
Lightbulb	P (W)	B (W/m^2)
1	2.4	0.19
2	3.6	0.29
3	4.8	0.38
4	7.2	0.57

12. If the student had tested a fifth lightbulb during Exercise 2 and measured the current passing through it to be 1.2 A, the Z associated with this bulb would have been:

F. 1 Ω.
G. 10 Ω.
H. 50 Ω.
J. 100 Ω.

GO ON TO THE NEXT PAGE.

13. Based on the results of Exercise 2, a circuit including the combination of which of the following batteries and lightbulbs would result in the highest current in the circuit? (Assume Z is a constant for a given lightbulb.)
    A. A 10 V battery and Bulb 1
    B. An 8 V battery and Bulb 2
    C. A 6 V battery and Bulb 3
    D. A 5 V battery and Bulb 4

14. With Bulb 3 in place in the outlet in Exercise 1, how many of the sensors were unable to detect any incident light?
    F. 1
    G. 2
    H. 3
    J. 4

15. Which of the following equations correctly calculates B (in W/m^2) at a distance of 2 m from Bulb 2?
    A. $B = \dfrac{2}{4\pi(3.6)^2}$
    B. $B = \dfrac{2}{4\pi(0.29)^2}$
    C. $B = \dfrac{3.6}{4\pi(2)^2}$
    D. $B = \dfrac{2.4}{4\pi(2)^2}$

16. Another student used the approach given in Exercise 3 to calculate B at a distance of 1 m from a fifth lightbulb. He determined that for this fifth bulb, B = 0.95 W/m^2. Accordingly, P for this bulb was most likely closest to which of the following values?
    F. 1 W
    G. 6 W
    H. 12 W
    J. 18 W

17. Which of the following was used as a method in Exercise 1 but NOT Exercise 2?
    A. Four light sensors were used.
    B. Four different lightbulbs were used.
    C. The electromotive force of the battery was varied.
    D. The current was highest for Bulb 4.

GO ON TO THE NEXT PAGE.

Passage IV

The electrons in a solid occupy energy states that are determined by the type and spatial distribution of the atoms in the solid. The probability that a given energy state will be occupied by an electron is given by the Fermi-Dirac distribution function, which depends on the material and the temperature of the solid.

Fermi-Dirac distribution functions for the same solid at three different temperatures are shown in the figure below.

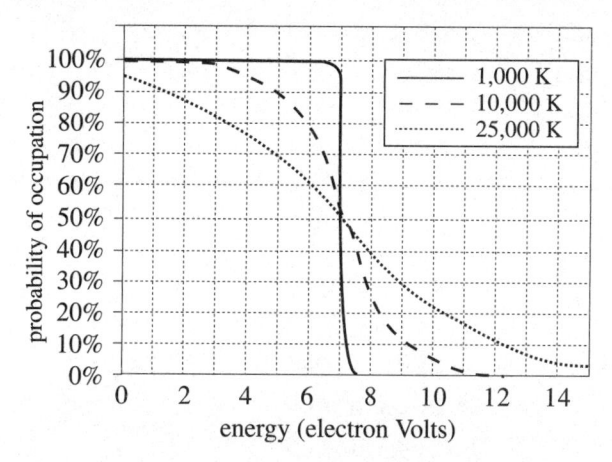

Note: 1 electron Volt (eV) = 1.66×10^{-19} Joules (J); eV and J are both units of energy. For energy states above 15eV, the probability of an electron occupying that state continues to decrease.

18. At the location where the probability of occupation equals 50%, the steepness of the slope of a distribution function is inversely proportional to the average *kinetic energy* (energy of motion) of the atoms in the solid. Which of the following correctly ranks the three functions according to the average kinetic energy of the atoms in the solid, from least to greatest?

   F. 25,000 K; 10,000 K; 1,000 K
   G. 1,000 K; 25,000 K; 10,000 K
   H. 10,000 K; 1,000 K; 25,000 K
   J. 1,000 K; 10,000 K; 25,000 K

19. Based on the figure, at a temperature of 1,000 K, the probability of a 20 eV energy state being occupied by an electron will most likely be:
   A. less than 5%.
   B. between 5% and 50%.
   C. between 50% and 90%.
   D. greater than 90%.

20. Based on the figure, which of the following sets of Fermi-Dirac distribution functions best represents an unknown solid at temperatures of 2,000 K, 20,000 K, and 50,000 K?

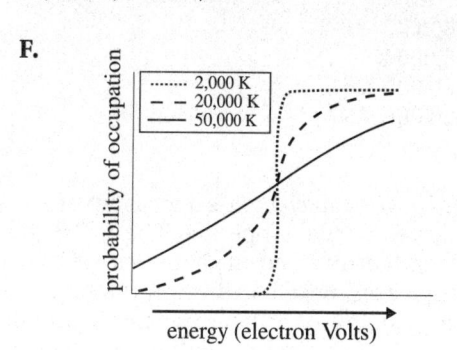

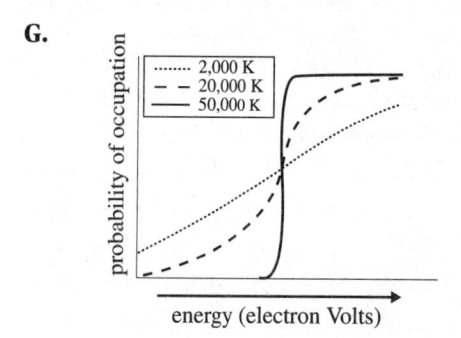

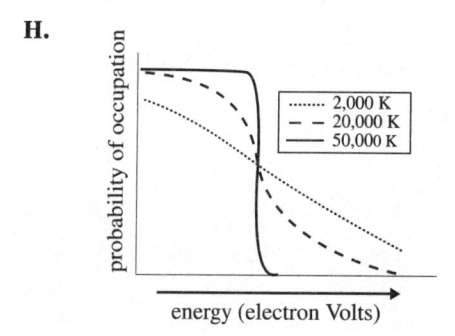

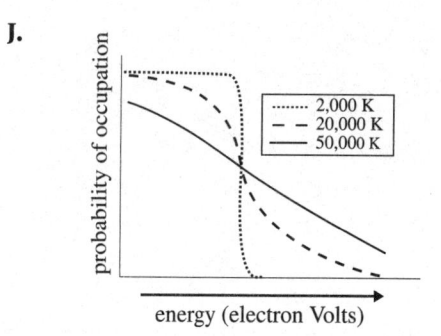

F.

G.

H.

J.

GO ON TO THE NEXT PAGE.

UNIT 1: GETTING TO KNOW THE ACT
LESSON A: PRACTICE TEST 1

21. Based on the figure, the probability of a 5 eV energy state being occupied by an electron will equal 80% when the temperature of the solid is closest to:
    A. 500 K.
    B. 5,000 K.
    C. 20,000 K.
    D. 30,000 K.

22. The *de Broglie wavelength* is a measure of a particle's wave-like properties. The de Broglie wavelength of an electron decreases as its energy state increases. Based on this information, over all energy states in the figure, as the de Broglie wavelength of an electron in that state decreases, the probability of that state being occupied by the electron:
    F. increases only.
    G. decreases only.
    H. increases, then decreases.
    J. decreases, then increases.

Passage V

A soda pop beverage is typically a solution of water, various liquid colorings and flavorings, and CO_2 gas. *Solubility* is defined as the ability of a substance to dissolve, and the solubility of CO_2 in a soda depends on the temperature and pressure of the system. As the temperature of a sealed container of soda changes, so does the solubility of the CO_2. This results in changes in the concentration of CO_2 in both the soda and the air in the container. The following experiments were performed to study the solubility of CO_2 in sodas.

Experiment 1

The apparatus shown in Figure 1 was assembled with a water bath at room temperature (25°C). After 10 minutes, the air pressure above the soda was measured in kilo-Pascals (kPa) by reading the value directly from the pressure gauge. Additional trials were performed at different temperatures and with other sodas in the container. The results are shown in Table 1.

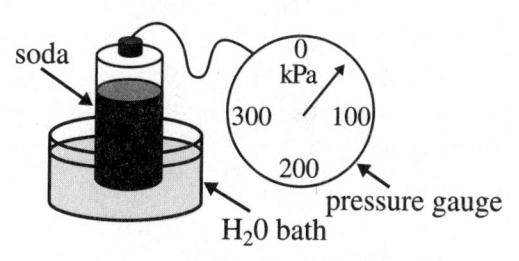

Figure 1

Table 1			
Soda	Pressure (kPa) at:		
	0°C	25°C	50°C
A	230	237	256
B	214	234	253
C	249	272	294
D	223	243	282
E	209	228	247

Experiment 2

An apparatus similar to those used by companies that produce soda was constructed, so that measured amounts of compressed CO_2 gas could be injected into each soda until the solution reached its maximum concentration of CO_2. The apparatus consisted of an air-sealed flask containing only soda and no air. Starting with sodas from which all of the CO_2 had been carefully removed, CO_2 was injected and the maximum CO_2 concentration for each soda was recorded. From the maximum concentrations, the solubility of CO_2 in each soda was calculated for three different temperatures at equal pressures. Solubility was recorded in centi-Molars per atmosphere (cM/atm), and the results are shown in Table 2.

Table 2			
Soda	CO_2 solubility (cM/atm) at:		
	0°C	25°C	50°C
A	3.59	3.51	3.45
B	3.58	3.50	3.37
C	3.67	3.57	3.48
D	3.62	3.53	3.41
E	3.54	3.46	3.29

23. Which of the following bar graphs best expresses the pressure of the container contents from Experiment 1 at 25°C?

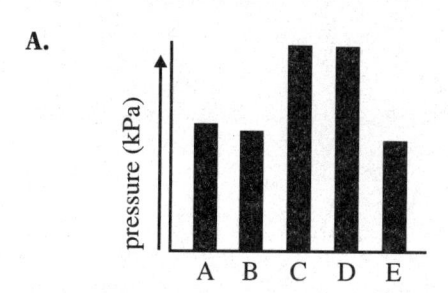

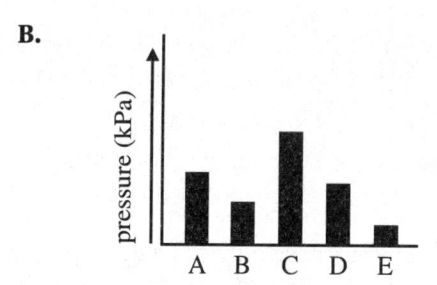

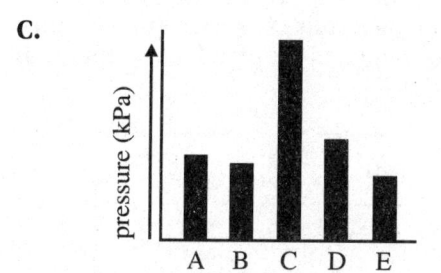

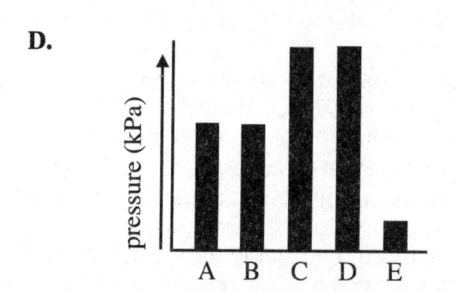

A.

B.

C.

D.

GO ON TO THE NEXT PAGE.

24. Which of the following figures best depicts the change in position of the needle on the pressure gauge, while attached to the container holding Soda C in Experiment 1?

   needle position at 0°C   needle position at 50°C
F.
G.
H.
J.

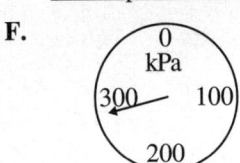

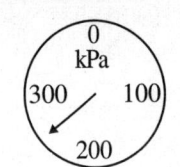

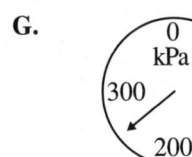

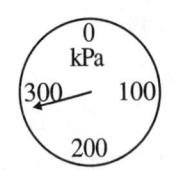

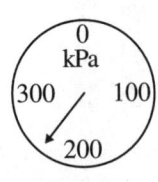

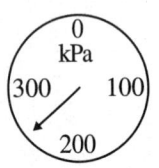

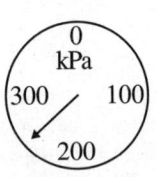

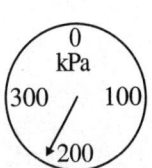

25. A student hypothesized that, at a given pressure and temperature, the higher the sugar content of a soda, the higher the solubility of CO_2 in that soda. Do the results of Experiment 2 and all of the information in the table below support this hypothesis?

Soda	Sugar content (grams per 12 ounces)
A	23
B	32
C	38
D	40
E	34

A. Yes; Soda A has the lowest sugar content and the lowest CO_2 solubility.
B. Yes; Soda D has a higher sugar content and CO_2 solubility than Soda C.
C. No; the higher a soda's sugar content, the lower the soda's CO_2 solubility.
D. No; there is no clear relationship in these data between sugar content and CO_2 solubility.

26. According to the results of Experiment 2, as the temperature of the soda increases, the CO_2 solubility of the soda:
F. increases only.
G. decreases only.
H. increases, then decreases.
J. decreases, then increases.

27. Which of the following figures best illustrates the apparatus used in Experiment 2?
A. 
B. 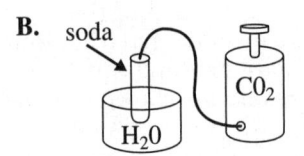
C. 
D.

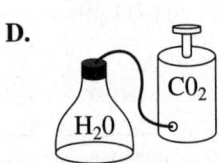

28. Which of the following statements best explains why in Experiment 1, the experimenter waited 10 minutes before recording the pressure of the air above the soda? The experimenter waited to allow:
F. all of the CO_2 to be removed from the container.
G. time for the soda in the container to evaporate.
H. the contents of the container to adjust to the temperature of the water bath.
J. time for the pressure gauge to stabilize.

GO ON TO THE NEXT PAGE.

Passage VI

Straight-chain conformational isomers are carbon compounds that differ only by rotation about one or more single carbon bonds. Essentially, these isomers represent the same compound in a slightly different position. One example of such an isomer is butane (C_4H_{10}), in which two methyl (CH_3) groups are each bonded to the main carbon chain. The straight-chain conformational isomers of butane are classified into four categories.

1. In the *anti* conformation, the bonds connecting the methyl groups to the main carbon chain are rotated 180° with respect to each other.

2. In the *gauche* conformation, the bonds connecting the methyl groups to the main carbon chain are rotated 60° with respect to each other.

3. In the *eclipsed* conformation, the bonds connecting the methyl groups to the main carbon chain are rotated 120° with respect to each other.

4. In the *totally eclipsed* conformation, the bonds connecting the methyl groups to the main carbon chain are parallel to each other.

The anti conformation has the lowest energy and is the most stable state of the butane molecule, since it allows for the methyl groups to maintain maximum separation from each other. The methyl groups are much closer to each other in the gauche conformation, but this still represents a relative minimum or metastable state, due to the relative orientations of the other hydrogen atoms in the molecule. Molecules in the anti or gauche conformations tend to maintain their shape. The eclipsed conformation represents a relative maximum energy state, while the totally eclipsed conformation is the highest energy state of all of butane's straight-chain conformational isomers. Two organic chemistry students discuss straight-chain conformational isomers.

Student 1

The *active shape* (the chemically functional shape) of a butane molecule is always identical to the molecule's lowest-energy shape. Any other shape would be unstable. Because the lowest-energy shape of a straight-chain conformational isomer of butane is the anti conformation, its active shape is always the anti conformation.

Student 2

The active shape of a butane molecule is dependent upon the energy state of the shape. However, a butane molecule's shape may also depend on temperature and its initial isomeric state. Specifically, in order to convert from the gauche conformation to the anti conformation, the molecule must pass through either the eclipsed or totally eclipsed conformation. If the molecule is not given enough energy to reach either of these states, its active shape will be the gauche conformation.

29. According to the passage, molecules in conformation states with relatively low energy tend to:

    A. convert to the totally eclipsed conformation.
    B. convert to the eclipsed conformation.
    C. maintain their shape.
    D. chemically react.

30. The information in the passage indicates that when a compound changes from one straight-chain conformational isomer to another, it still retains its original:

    F. energy state.
    G. shape.
    H. number of single carbon bonds.
    J. temperature.

GO ON TO THE NEXT PAGE.

31. Student 2's views differ from Student 1's views in that only Student 2 believes that a butane molecule's active shape is partially determined by its:
    A. initial isomeric state.
    B. energy state.
    C. hydrogen bonding angles.
    D. proximity of methyl groups.

32. Which of the following statements would Student 1 and Student 2 both agree with?
    F. The totally eclipsed state will sometimes be the active state.
    G. The gauche state will never be the active state.
    H. If the molecule starts in anti, that will be its active state.
    J. Eclipsed is the most likely state to be active.

33. Suppose butane molecules are cooled, and the active shape of each molecule is examined. Which of the following statements is most consistent with the information presented in the passage?
    A. If Student 1 is correct, all of the molecules will be in the anti conformation.
    B. If Student 1 is correct, all of the molecules will have shapes different from their lowest-energy shapes.
    C. If Student 2 is correct, all of the molecules will be in the anti conformation.
    D. If Student 2 is correct, all of the molecules will have shapes different than their lowest-energy shapes.

34. Which of the following diagrams showing the relationship between a given butane molecule's shape and its relative energy is consistent with Student 2's assertions about the energy of butane molecules, but is NOT consistent with Student 1's assertions about the energy of butane molecules?

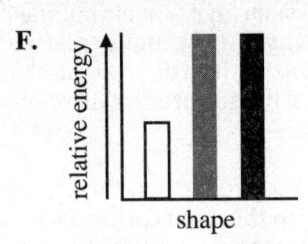

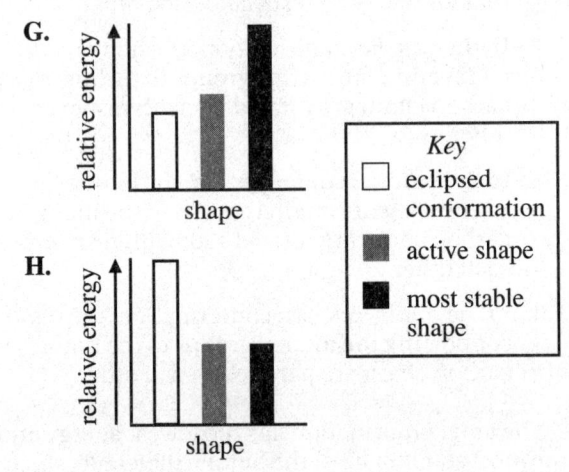

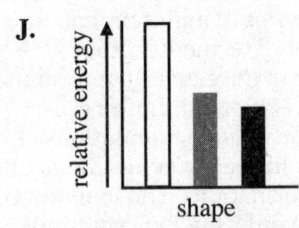

    F.
    G.
    H.
    J.

    Key
    □ eclipsed conformation
    ▨ active shape
    ■ most stable shape

35. Student 2 says that a butane molecule may settle into a moderately high-energy conformation. Which of the following findings, if true, could be used to counter this argument?
    A. Once a molecule has settled into a given conformation, all of its single carbon bonds are stable.
    B. Enough energy is available in the environment to overcome local energy barriers, driving the molecule into its lowest-energy conformation.
    C. During molecule formation, the hydrogen bonds are formed before the carbon bonds.
    D. Molecules that change their isomeric conformation tend to lose their chemical functions.

GO ON TO THE NEXT PAGE.

Passage VII

The survival of plant life depends heavily on the availability of nitrogen in the environment. Although about 72% of Earth's atmosphere consists of N_2 gas, this form of nitrogen is inaccessible to plants, because a plant cell is incapable of breaking the triple bond between the two nitrogen atoms. Certain bacteria in soil, however, are capable of processing N_2 into ammonia (NH_3), a form of nitrogen that plants can utilize. This process is called *nitrogen fixation*. Plant roots extract nitrogen in the form of ammonia from the soil and release back into the soil various forms of nitrogen as metabolic byproducts. After a plant dies, it also releases various forms of nitrogen as it decays. Figure 1 shows how the concentration of ammonia in the soil affects the growth rate of a certain bean plant. Figure 2 shows the typical ammonia concentrations found in soil at various depths beneath the surface. Figure 3 shows the concentrations of two different forms of nitrogen found at equal depths beneath the soil surface in three different environments. The concentrations in all three figures are given in units of parts per million (ppm).

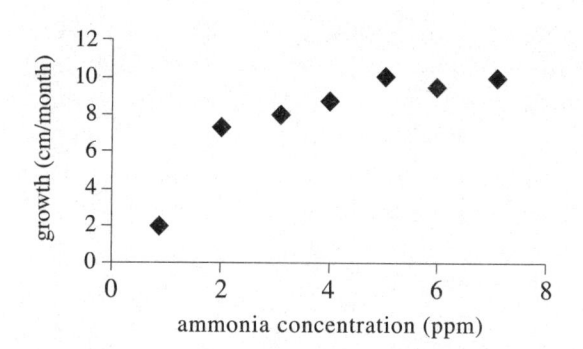

Figure 1

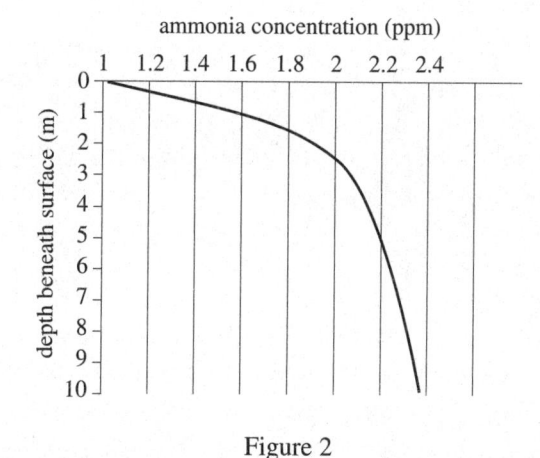

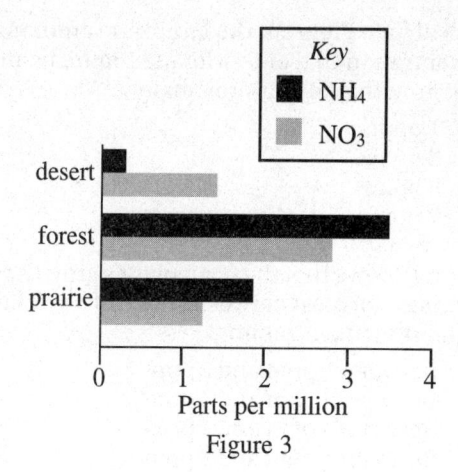

Figure 2

Figure 3

36. Assume that the soil samples in Figure 3 were extracted from a depth of 2 m beneath the soil surface. Based on the information in Figures 2 and 3, the soil samples analyzed in Figure 2 were most likely taken from which environment?

    F. Desert
    G. Forest
    H. Prairie
    J. The environment from which the samples were taken cannot be determined from the information given in Figures 2 and 3 alone.

37. The data in Figure 3 supports which of the following statements about nitrogen fixation?

    A. Forest bacteria are incapable of nitrogen fixation.
    B. Desert bacteria are incapable of nitrogen fixation.
    C. Bacteria capable of nitrogen fixation are much more prevalent in a forest environment than in a desert environment.
    D. Bacteria capable of nitrogen fixation are much less prevalent in a forest environment than in a desert environment.

38. Bean plants at higher altitudes tend to grow faster than those at lower altitudes. According to Figure 1, this most likely occurs because the soil at higher altitudes:

    F. supports fewer bacteria capable of nitrogen fixation.
    G. has a higher concentration of forms of nitrogen other than ammonia.
    H. has a lower ammonia concentration.
    J. has a higher ammonia concentration.

GO ON TO THE NEXT PAGE.

39. According to Figure 1, the minimum ammonia concentration that allows for maximum bean plant growth rate is approximately:

- **A.** 1 ppm.
- **B.** 2 ppm.
- **C.** 5 ppm.
- **D.** 7 ppm.

40. Figure 1 shows that the bean plant's growth rate increases the most between which of the following ammonia concentrations?

- **F.** Between 1 ppm and 2 ppm
- **G.** Between 2 ppm and 3 ppm
- **H.** Between 3 ppm and 4 ppm
- **J.** Between 4 ppm and 5 ppm

END OF TEST.

STOP! DO NOT TURN THE PAGE UNTIL TOLD TO DO SO.

lesson B
The Landscape of the Test

Thinking KAP

You entered a contest with a travel agency and you won a trip for four to any country in the world. Your parents have given you and your three best friends permission to go on your own.

Together, you and your friends need to decide where you want to go. What are the first steps you would take to prepare for the journey?

Strategy Instruction

Know the Test

In order to prepare for a journey, it is essential to first know some basic facts. Where are you going? How long will you stay? What kinds of things will you do? Similarly, in order to prepare for the Science ACT, you should learn as much as you can about the test itself. Once you know the landscape of the test, you can better prepare strategies to navigate your way to a successful conclusion.

Timing and the Science ACT

The Science ACT has 40 questions which are related to 7 passages, which you must answer in 35 minutes. This gives you about 5 minutes to read each passage and answer 5–7 questions afterwards. That sounds impossible! How will you do it?

First, you need to know the ACT. By the end of this course, you will know which questions you can answer easily, which ones require an educated guess, and which ones to skip. You will also know when to read the text in detail, and when to skim the text for correct answers.

Remember, each question is worth one point. If you spend 5 minutes on a hard question and run out of time for 5 easy questions, you've just lost 5 points. It's much better to make an educated guess and move on—there's a good chance you'll be right!

Look back at your practice test. What did you learn about the ACT? What will you do differently next time?

_____
_____
_____
_____

keep in mind

The better you understand the Practice Test, the better you'll be prepared for the real ACT.

The 3-Step Method for Science Passages

A key method for using your time efficiently is to answer the questions in the right order. This keeps you from wasting your time reading material you don't need and prevents you from getting confused about where to find information. The following is a basic 3-Step Method for Science Passages. As you learn more about the types of questions on the ACT, you will learn different variations of this basic 3-Step Method.

STEP 1: Read the introduction.

The introduction tells you what the passage is about, defines vocabulary for the passage, and may contain the answers to some questions. Since it is only 1–4 sentences long, reading it first is a good use of your time.

STEP 2: Answer the easy questions.

Easy questions address *only one* experiment, viewpoint, table, or figure. Some easy questions can be answered with just the introduction.

STEP 3: Answer the harder questions.

Harder questions address *more than one* experiment, viewpoint, table, or figure.

keep in mind

Once you answer all the easy questions for a passage, you will have the information you need to answer the harder questions.

Read What You Need

Did you know that you can answer most of the questions for some passages without reading all of the text? Most questions can be answered directly from the graphs, or using just one or two sentences from the text.

Examine Passage V of the Practice Test. Read each question, and then determine where the information is found.

Question	Which data source did you use to find the answer?
23	*Table 1 only*
24	
25	
26	
27	
28	

Did you need to read all of the text in the passage? _____

Did you need to understand the experiments to answer questions about the tables? _____

Which types of questions require you to read the text? _____
_____

keep in mind

In Unit 2, you'll learn how to use keywords to read only the text you need.

Finding the Easy Ones

To find an easy question, look for the following:

- the title of an experiment, viewpoint, table, or figure
- the words "according to the passage"

Go back to the Practice Test at the beginning of this unit. Work independently or with a partner to find the easy questions and the harder questions.

Passage Type	Easier Questions	Harder Questions
Passage I		
Passage II		
Passage III		
Passage IV		
Passage V		
Passage VI		
Passage VII		

keep in mind

If you know where the easy questions are, you can build points faster.

Are the easy questions at the beginning, at the end, or mixed around? How will this help you find them?

_____

_____

Guided Practice

Use the 3-Step Method for Science Questions to read the passage and answer the questions.

The ash produced by an erupting volcano can sometimes be more harmful than the lava itself. Lava moves slowly and predictably and can be avoided with sufficient warning; however, the ash can be deposited over a distance of many miles over a period of several years, causing numerous health and environmental problems. Some of these problems are directly related to the particle size of the ash. In particular, very small particulates may travel deeply into a person's lungs, causing asthma, emphysema, and other related diseases. Large quantities of ash, in contrast, can suffocate plants and other ground-dwelling life.

Experiment 1

Scientists hypothesized that particle diameter would affect how far particles traveled after a volcanic eruption. They believed that smaller particles would be more likely to stay in the air longer, and thus travel a greater distance. In order to test their theory, they determined the mean diameter of ash particles that fell to the ground downwind of Mount St. Helens during the 8-hour-long eruption on May 18, 1980.

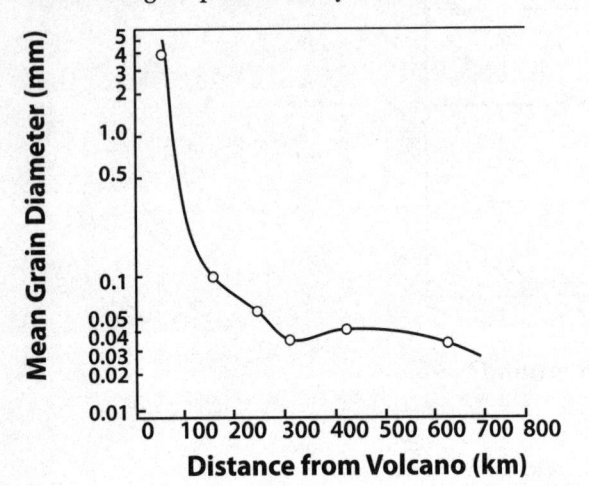

Figure 1

Experiment 2

Scientists hypothesized that the thickness of the ash layer would depend on numerous factors, including volume of ash released, particle size, height and duration of the eruption, and prevailing wind conditions. Overall, they expected to see an exponential decrease in the thickness of the volcanic ash layer as they moved farther away from the eruption. The data they collected from the 1980 Mount St. Helens eruption is shown in Figure 2.

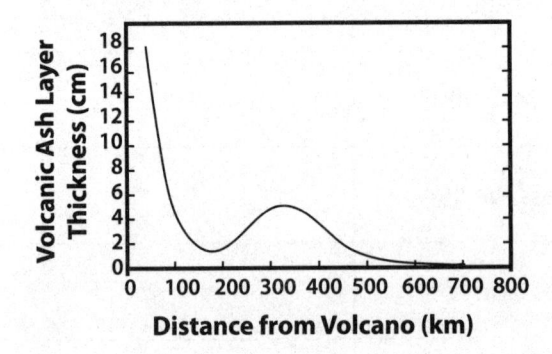

Figure 2

1. How thick was the ash layer at the location where the average particle diameter was 0.1 mm?
   A. 2 cm
   B. 4 cm
   C. 6 cm
   D. 8 cm

What information did you need to answer this question? _____

Is this an "easy" or a "hard" question? _____

2. The thickness of the ash deposits in the Mount St. Helens eruption did not follow the expected pattern. Scientists believed that this was due to local humidity levels causing the particles to clump together in the area approximately 300 km downwind of the volcano. According to Figure 1, what was the approximate diameter of the individual ash particles forming those clumps?
   F. 0.04 mm
   G. 0.05 cm
   H. 4 cm
   J. 5 cm

What information did you need to answer this question? _____

Is this an "easy" or a "hard" question? _____

Shared Practice

Work with a partner to complete the section below. Remember to answer the easy questions first, and only read what you need.

Mountain Gorillas are an endangered species native to the mountains of Rwanda, Uganda, and Zaire. They live in troops of up to 30 individuals. Although scientists have studied Mountain Gorilla troops in captivity for years, they do not fully understand the factors affecting their growth and survival in the wild. Recently, scientists attempted to develop a computer model that, based on data gathered from troops living in nature preserves, would predict the growth and success of a troop in the wild.

Experiment 1

Scientists obtained data on the deaths of members of gorilla troops in three different nature preserves. They recorded the percentage of gorillas surviving at each age. This is shown in Figure 1.

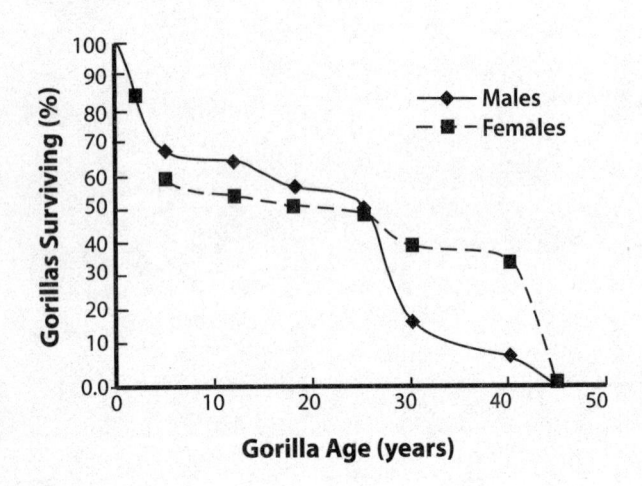

Figure 1

Experiment 2

To expand the model, scientists examined additional factors that affected population growth, including information on the probability of death at any given age, social structure, birth rate, age at first pregnancy, and time elapsed between pregnancies. Using the model, they created a computer simulation of a population of gorillas that mimicked the gorilla troops used in their research. They then looked at how well the model predicted various factors that affect the size and growth of the gorilla population. Table 1 shows the data on variables related to birth and death.

Table 1	Computer Model	Research Results
Birth and Death Rates		
Deaths per gorilla per year	0.081	0.073
Births per gorilla per year	0.052	0.056
Births per adult female per year	0.253	0.27
Infanticide		
Percent of infant mortality	28%	25%
Percent of all births	8%	11%
Average age of first pregnancy		
Groups with one male	10.9 years	11.1 years
Groups with more than one male	9.6 years	9.9 years
Average time between births		
When previous offspring survives	3.9 years	4.0 years
When previous offspring dies	2.3 years	2.2 years

1. The size of gorilla troops is determined by numerous factors. The scientists involved in Experiment 2 would predict that all of the following affect the growth of gorilla troops except:
   A. the number of females giving birth in a year.
   B. the number of adult males in the troop.
   C. whether a female's previous offspring was male or female.
   D. the number of adult females in the troop.

2. Which of the following statements is supported by Figure 1?
   F. The death rates of males in the wild cannot be explained by the model.
   G. Female gorillas are significantly more likely than male gorillas to die before the age of 40 years.
   H. For any age, the number of female gorillas surviving is greater than the number of male gorillas surviving.
   J. Male gorillas are significantly more likely than female gorillas to die after the age of 30 years.

3. Percent error is a measure of the accuracy of any measured value. It is determined by comparing the difference between the actual value and the predicted value to the actual value itself. Percent error is expressed mathematically as

   $$\text{percent error} = 100 \times \frac{|\text{actual value} - \text{predicted value}|}{\text{actual value}}$$

   What is the percent error of the measurement for infanticide as a percentage of all births?
   A. 3%
   B. 10%
   C. 27%
   D. 38%

4. Which of the following statements is supported by the data?
   F. The percentage of female gorillas in a troop becomes larger than the percentage of male gorillas at approximately the age when they begin to give birth.
   G. The growth rate for a troop will be higher when infant mortality is higher.
   H. As the birth rate for a gorilla troop decreases, its death rate increases.
   J. The average age of first pregnancy for a female gorilla is lower when there is more than one male in a troop.

5. In Table 1, the number of births per gorilla per year is much smaller than the number of births per adult female per year. What is the best explanation for this discrepancy?
   A. A large proportion of gorillas in a troop are males and juvenile females.
   B. A large proportion of gorillas in a troop are females of all ages.
   C. A large proportion of gorillas in a troop are males, juvenile females, and females who have recently given birth.
   D. A large proportion of gorillas in a troop are adult females and the infants they have just given birth to.

Name _____ Date _____

KAP Wrap

Write a letter to your teacher, telling him or her your thoughts about the ACT Science Test. In your letter, be sure to say:

- what areas you are strong in.
- what areas you think you need to work on.
- what you think you can do to get ready for the test.
- how you think your teacher can help you get ready for the test.

lesson C Eliminating

ReKAP

Review the strategies from this unit. Then fill in the blanks with what you have learned.

1. The Science ACT contains _____ questions, which address _____ passages. These questions must be answered in _____ minutes.

2. Easy questions will often contain the name of *one* _____, _____, _____, or _____. Easy questions that can be answered by the introduction may contain the words "_____."

3. When approaching a passage, you should read the entire _____, but only read the rest of the text if you need to.

4. Harder questions will ask about _____.

Answer true or false for the questions below:

5. Harder questions are worth more points. _____

6. If you haven't studied a topic in class, you can still do well on that passage. _____

7. It's better to leave an answer blank than to guess. _____

8. You need to read every word and understand the data before you can answer the questions. _____

Strategy Instruction

Eliminating

One of the most powerful strategies you can use on the ACT is Eliminating wrong answers. There is no penalty for guessing, so you want to fill in every bubble and make your guesses count!

A random guess will give you a 25% chance of getting the correct answer. If you can eliminate two wrong answers, you have a 50% chance. And of course if you can eliminate three wrong answers, you've solved the problem.

keep in mind

When eliminating a wrong answer, draw a line through it—you are less likely to second-guess yourself.

Eliminating

- Eliminate answers that you know are wrong.
- Guess the answer that looks best from the ones that are left.

Try It Out!

Underline keywords in the passage below; then use Eliminating to answer the questions on the next page.

Passage I

When the *Voyager 2* spacecraft passed near Triton, the largest moon of Neptune, it sent back pictures of a complex network of ridges and valleys partially covered with a thick layer of methane and nitrogen "ice." Two views on the causes of surface structure formation on Triton are presented below.

Scientist 1

The images of Triton seem to indicate recent volcanic activity, probably as recently as millions or tens of millions of years ago. There are many features clearly visible which appear to be recently-formed lava lakes, similar to the ice-lava lakes found on Jupiter's moon Ganymede. Other features resemble the icy lava fracture flows on Uranus' Ariel. There even seems to be ongoing volcanic activity, explaining the dark streaks in several places on the surface. Though Triton is a small planetary body, it is still possible that it has a source of internal heat, similar to the radioactive decay of Earth, which supplies sufficient energy for volcanic reactions. Only a small amount of internal heat would be necessary to create a nitrogen volcano.

Scientist 2

The landforms on Triton are caused by nitrogen precipitation and glacial movement. The lakes and flow forms are the result of glacier-like movements of the methane-nitrogen polar ice caps. The melting and freezing of these caps is caused by seasonal heating by the Sun. Dark streaks seen on Triton's surface result from methane-nitrogen "snow fall" during the radical season changes which occur as Triton's poles take turns facing the sun.

1. Which of the following discoveries, if true, would most conclusively support Scientist 1?

   A. The surface of Triton has the same composition as Ganymede and Ariel.
   B. The apparent small craters seen in the high-resolution pictures of Triton were introduced by faulty image processing.
   C. Further inspection of images from Triton reveals an erupting volcano.
   D. Triton is shown not to have been captured, but to have originated together with Neptune.

2. New research suggests that slow radioactive decay may persist in a body much smaller than Earth for up to 10 billion years. This information, if true, best supports the viewpoint of:

   F. Scientist 1.
   G. Scientist 2.
   H. both Scientist 1 and Scientist 2.
   J. neither Scientist 1 nor Scientist 2.

> **keep in mind**
>
> All questions about why an author mentioned something have the same answer: it supports the hypothesis.

UNIT 1: GETTING TO KNOW THE ACT
LESSON C: ELIMINATING

Name_____ Date_____

Test Practice Unit 1

When your teacher tells you, carefully tear out this page. Then begin working.

1. Ⓐ Ⓑ Ⓒ Ⓓ   15. Ⓐ Ⓑ Ⓒ Ⓓ

2. Ⓕ Ⓖ Ⓗ Ⓙ   16. Ⓕ Ⓖ Ⓗ Ⓙ

3. Ⓐ Ⓑ Ⓒ Ⓓ

4. Ⓕ Ⓖ Ⓗ Ⓙ

5. Ⓐ Ⓑ Ⓒ Ⓓ

6. Ⓕ Ⓖ Ⓗ Ⓙ

7. Ⓐ Ⓑ Ⓒ Ⓓ

8. Ⓕ Ⓖ Ⓗ Ⓙ

9. Ⓐ Ⓑ Ⓒ Ⓓ

10. Ⓕ Ⓖ Ⓗ Ⓙ

11. Ⓐ Ⓑ Ⓒ Ⓓ

12. Ⓕ Ⓖ Ⓗ Ⓙ

13. Ⓐ Ⓑ Ⓒ Ⓓ

14. Ⓕ Ⓖ Ⓗ Ⓙ

UNIT 1: GETTING TO KNOW THE ACT
LESSON C: ELIMINATING

SCIENCE TEST
15 Minutes—16 Questions

DIRECTIONS: There are three passages in this test. Each passage is followed by several questions. After reading a passage, choose the best answer to each question and fill in the corresponding oval on your answer sheet. You may refer to the passages as often as necessary.

You are NOT permitted to use a calculator on this test.

Passage I

Water is the basis of all life on Earth and a powerful solvent which can dissolve a wide variety of substances. Water exists in three physical states: as solid ice, liquid water, and gaseous vapor. In solid ice, many complex variations have been observed in the networks formed by its molecules.

Table 1 describes the properties of some of the known molecular arrangements, or polymorphs, of ice.

Table 1			
Polymorph	Crystalline Network Shape	H_2O Molecules per Unit Crystal Cell	Density (g/cm³) at 10^6 Pa
Hexagonal ice (Ih)	Hexagonal	6	0.91
Cubic ice (Ic)	Cubic	8	0.92
Ice-two (II)	Rhombohedral	12	1.17
Ice-three (III)	Tetragonal	12	1.16
Ice-five (V)	Monoclinic	28	1.24

The physical state of water is determined by several different factors. Phase diagrams illustrate the effects of these factors. Figure 1 shows the conditions at which water is a gas, liquid, or solid, in several polymorphs.

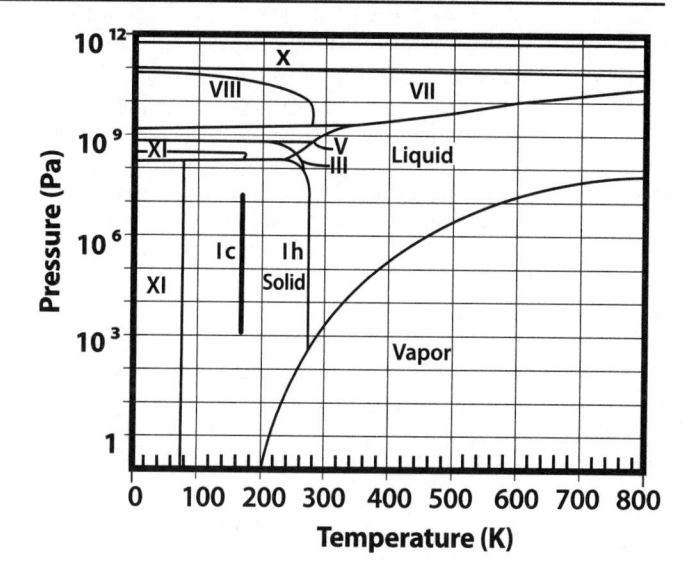

Figure 1

1. The density of liquid water is defined as 1.00 g/cm³ when the pressure is 10^6 Pa. According to Table 1, at this pressure, the polymorphs of ice:

   A. are always more dense than liquid water.
   B. are always less dense than liquid water.
   C. are the same density as liquid water.
   D. can be more dense or less dense than liquid water.

2. Surface conditions on Earth are in the range of approximately 203 K to 323 K. At sea level, where pressure is 10^6 Pa, which polymorph of ice does Figure 1 indicate is most likely to form on Earth's surface?

   F. Ih
   G. Ic
   H. VII
   J. X

GO ON TO THE NEXT PAGE.

3. Sublimation is defined as a phase change directly from a solid to a gas. According to Figure 1, under which of the following conditions of temperature and pressure would sublimation of ice occur?
   A. T = 353 Kelvin, P = 10^4 Pa.
   B. T = 253 Kelvin, P = 10^4 Pa.
   C. T = 353 Kelvin, P = 10^2 Pa.
   D. T = 253 Kelvin, P = 10^2 Pa.

4. Which of the following hypotheses best summarizes the relationship between the number of molecules in a unit crystal cell and the density of polymorphs, as described in Table 1?
   F. In general, the higher the number of molecules in the unit crystal, the higher the density of the polymorph.
   G. In general, the higher the number of molecules in the unit crystal, the lower the density of the polymorph.
   H. The number of molecules in the unit crystal varies, while the density of the polymorph is constant.
   J. No relationship exists between the number of molecules in unit crystals and the density of the polymorph.

5. The triple point defines the conditions of temperature and pressure at which all three phases of matter coexist, though they will rapidly change to one defined state if the temperature or pressure changes even slightly. According to Figure 1, what is the approximate pressure at the triple point of gaseous, liquid, and solid water?
   A. 10^7 Pa
   B. 10^5 Pa
   C. 10^3 Pa
   D. 10 Pa

6. Based on the information in Figure 1, at 650 K and 10^9 Pa, in which form does H_2O occur?
   F. Ice VII
   G. Ice XI
   H. Liquid water
   J. Gaseous vapor

GO ON TO THE NEXT PAGE.

Passage II

Phytoplankton are simple unicellular organisms that use photosynthesis to create a large portion of the breathable oxygen in the atmosphere and the dissolved oxygen in the oceans. They are also the primary food source for zooplankton (unicellular animal-like organisms) and tiny fish, which are in turn the food sources for larger fish.

Phytoplankton levels are measured by obtaining a sample of ocean water and testing for fluorescence from chlorophyll-a, the only type of chlorophyll that all types of phytoplankton possess. Phytoplankton are sensitive to several environmental factors, such as temperature and salinity (concentration of salts), which in turn are influenced by seasonal changes.

Oceanographers measured the temperature, salinity, and chlorophyll-a fluorescence at different depths, on single days in Fall, Winter, and Summer, for a bay near Australia. Figure 1 illustrates their results.

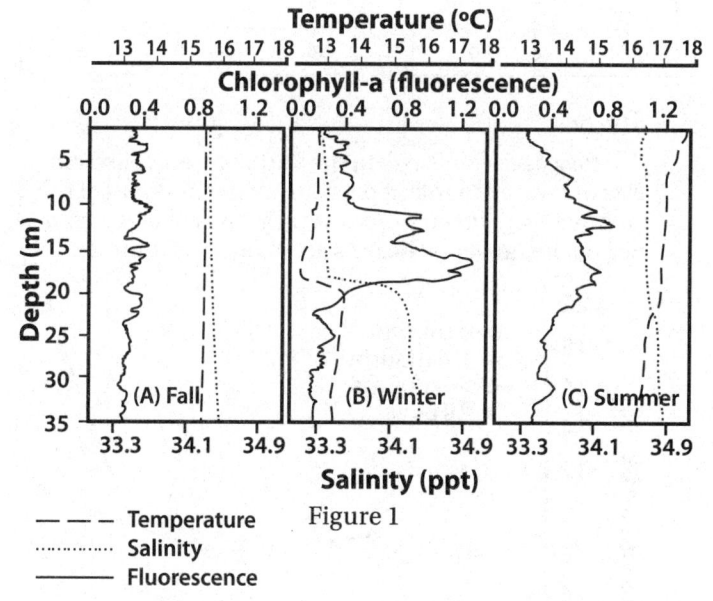

Figure 1

- - - Temperature
.... Salinity
——— Fluorescence

7. According to Figure 1, which of the following conditions is the least affected by seasonal change?
   A. Surface water temperature
   B. Deep water temperature
   C. Surface water salinity
   D. Deep water salinity

8. According to Figure 1, during the summer the largest concentration of phytoplankton is at approximately what depth?
   F. 5 m
   G. 15 m
   H. 25 m
   J. 35 m

9. Studies indicate that surface water temperatures are coldest in the winter and warmest in the summer. Given this information and the data summarized in Figure 1, what can you conclude is the minimum temperature range of surface water, over the course of the year?
   A. 1°C
   B. 3°C
   C. 5°C
   D. 10°C

10. Temperature inversion occurs when deep waters have a higher temperature than surface waters, causing the deeper, warm waters to move upward, while the cool water at the surface sinks. According to Figure 1(B), what effect does temperature inversion have on chlorophyll-a fluorescence?
    F. Temperature inversion does not cause any change in chlorophyll-a fluorescence.
    G. Temperature inversion causes chlorophyll-a fluorescence to increase at all depths.
    H. Temperature inversion causes chlorophyll-a fluorescence to decrease at all depths.
    J. Temperature inversion causes chlorophyll-a fluorescence to increase at intermediate depths where the water turns over.

11. Which of the following statements best describes the relationship between salinity and the other data in Figure 1?
    A. As temperature increases, salinity increases.
    B. As temperature increases, salinity decreases.
    C. As chlorophyll-a fluoroscence increases, salinity increases.
    D. As chlorophyll-a fluoroscence increases, salinity decreases.

GO ON TO THE NEXT PAGE.

Passage III

Antibiotics are medications which function by killing bacteria that live within other living organisms. In veterinary medicine, there are several groups of antibiotics, used on animals, which are effective in treating a variety of different bacterial infections.

Antibiotic resistance occurs when an antibiotic is given and one or more bacteria happen to have a DNA mutation that prevents them from dying. These bacteria reproduce while the remaining bacteria die. As a result, the host organism will carry an entire population of bacteria that have a genetic mutation that makes them resistant to the antibiotic.

Study 1

Escherichia coli (*E. coli*) lives in the intestines of warm-blooded organisms and helps in the digestion of food. However, there are some strains of *E. coli* which produce toxins that can cause severe illness or even death. These strains can be passed from livestock, such as pigs, to humans who consume them. Figure 1 shows the rates of *E. coli* resistance in pigs to several antibiotics over a 5-year period.

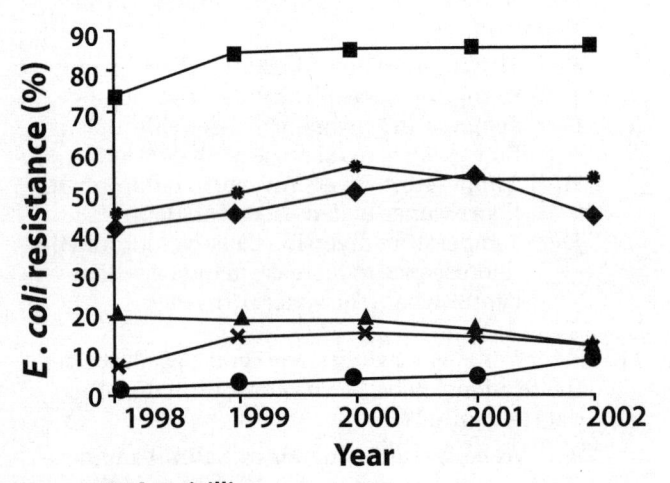

Study 2

E. coli is used as a standard for studies with antibiotics. However, many other bacteria are far more harmful to pigs and to human consumers. In Table 1, four strains of bacteria that cause infection in pigs were compared for their rates of resistance to several antibiotics.

Table 1				
Antibiotic	Percent Resistance to Antibiotic			
	P. multocida	*A. pleuropneumoniae*	*S. suis*	*A. pyrogenes*
Ampicillin	4	3	0	0
Penicillin	0	0	0	0
Tetracycline	9	25	92	0
Trimethoprim/ sulphonamide	7	13	6	10
Enrofloxacin	0	4	0	0
Ceftiofur	0	0	0	0

Study 3

Resistance of *E. coli* to the antibiotic enrofloxacin was measured in infant pigs during the first year of life. Samples were taken from several groups of pigs over a period of four years. Results are shown in Figure 2.

Figure 1

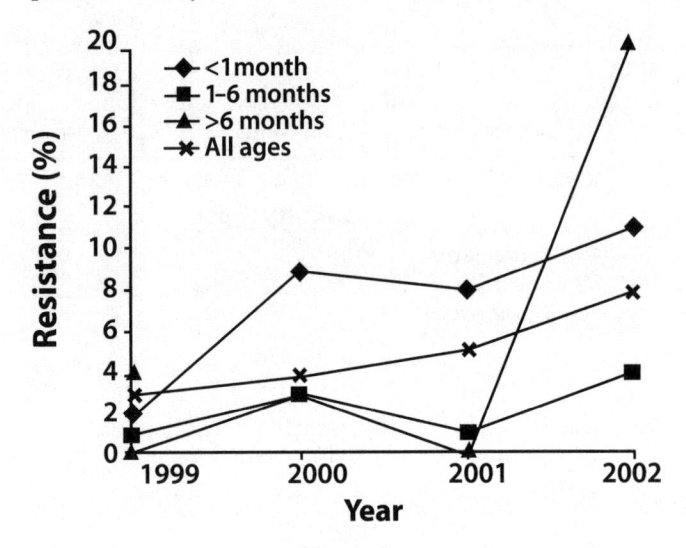

Figure 2

12. Scientists recommend that antibiotics be used only when absolutely necessary. If an antibiotic is not used as frequently, the resistance to it may decrease over time. For which antibiotic does *E. coli* show the greatest decrease in resistance from 1998 to 2002, shown by Study 1?
    F. Ampicillin
    G. Tetracycline
    H. Neomycin
    J. Apramycin

13. How much has *E. coli* resistance to enrofloxacin, in pigs of all ages, increased from the year 2000 to 2002, as depicted in Figure 2?
    A. 1%
    B. 3%
    C. 7%
    D. 10%

14. According to Figure 2, which age group shows the lowest percentage of enrofloxacin resistance for *E. coli*, in 2001?
    F. Pigs under 1 month old
    G. Pigs between 1 and 6 months old
    H. Pigs over 6 months old
    J. Pigs of all ages

15. Studies show that those antibiotics that are in highest use have the greatest risk of developing resistance from bacteria. Based on this information and the data in Studies 1 and 2, which antibiotic is most widely used in pigs?
    A. Ampicillin
    B. Penicillin
    C. Tetracycline
    D. Enrofloxacin

16. According to Table 1, which antibiotic is likely to produce resistant strains in the greatest number of bacterial species?
    F. Ampicillin
    G. Enrofloxacin
    H. Penicillin
    J. Trimethoprim/sulphonamide

END OF TEST.

STOP! DO NOT TURN THE PAGE UNTIL TOLD TO DO SO.

KAP Wrap

You now have a good idea about the structure and scoring of the ACT. If you could retake the Practice Test, what would you do differently?

Unit 2
Data Representation Passages

lesson A: Find the Keywords

Thinking KAP

You and your friends have decided to go to Mexico. There are miles of beautiful beaches, you will be able to explore a new culture, you will see plenty of wildlife, and you will get to practice your Spanish. But what will you do once you are there? The library, the bookstore, and the Internet all feature a wealth of resources, but there are too many options to browse them all. You do not want to get guidebooks that are not helpful, or read information that is not useful to you. What techniques can you use to find the information you need?

Strategy Instruction

Find the Data

Although science questions on the ACT may be about almost any topic, from car exhaust to oceanic shrimp, they are actually testing only a few skills. Fortunately, these are skills that you use every day when looking up information on the Internet, in a magazine, or in a bookstore.

Let's say you want to find good vacation spots in Mexico. First, you might go to an Internet search engine and type in a few *keywords*, like "Mexico" and "vacation." To decide what you want to read, you might *skim the titles* and *select a page*. Of course, the page might contain information you don't need, so you would *look for headings* to find the information. Finally, you would *read just the information you need*. You don't have to know anything about Mexico to do this search. And you don't need to know anything about oceanic shrimp to take the ACT.

In this lesson, you will learn how to think through common ACT questions before you answer them. The 3-Step Method for Data Representation will help you answer questions in which you are presented with an array of data, multiple sources of information, and figures such as charts, tables, and diagrams.

The 3-Step Method for Data Representation

STEP 1: Highlight keywords.

STEP 2: Use the diagrams.

STEP 3: Answer the question.

> **keep in mind**
>
> This process can help you do better on every single question. Even if a question seems easy, it will prevent you from making mistakes.

Keywords Tell You What You Know

All questions have two sets of keywords. If you can locate both sets, you can often answer the question without even reading the passage above the diagram.

> **Highlighting Keywords**
>
> - **What do you know?**
>   Underline the keywords that tell you what you know.
>
> - **What do you need?**
>   Circle the keywords that tell you what you need to find out.

What do you know?

"What you know" keywords tell you three things:

- the name of the diagram.
- the data point or trend that is known.
- the title of the column or axis for the known data point or trend.

keep in mind

"What you know" keywords are not definitions or explanations of how an experiment works. They are words that provide you with information about useful results.

Try It Out!

Skim the introductory paragraph below. Then underline the "What you know" keywords for both questions.

Two scientists are conducting studies on the behavior of gases. The first is experimenting with the relationship between temperature and volume. The second is experimenting with the relationship between pressure and temperature.

1. According to Study 2, in which the volume of the container is constrained to 20 gallons, if the pressure in the chamber increases from 760 mm Hg to 950 mm Hg, then it can be assumed that the temperature would:

2. The scientist in Study 1 brings the chamber to a temperature of 450°F. Assuming the relationship between temperature and volume is consistent, to which of the following will the volume be closest?

UNIT 2: DATA REPRESENTATION PASSAGES
LESSON A: FIND THE KEYWORDS

Keywords Tell You What You Need

"What you need" keywords help you focus on the information you need and where it can be found.

What do you need?

"What you need" keywords tell you two important things:

- the label of a specific column, row, or axis in which the information can be found.
- if the information you need is a data point or a trend.

Sometimes, there will not be a keyword directly telling you whether you are looking for a data point or for a trend. However, if the "What you know" information is a data point, you are probably looking for a data point. Similarly, if the "What you know" information is a trend, you are probably looking for a trend.

keep in mind

Often, if the "What you know" information describes a row, the "What you need" information will describe a column. The data point where they meet will often be the answer.

Try It Out!

Underline the "What you need" keywords for both questions, and state whether you are looking for a data point or a trend.

Two scientists are conducting studies on the behavior of gases. The first is experimenting with the relationship between temperature and volume. The second is experimenting with the relationship between pressure and temperature.

1. According to Study 2, in which the volume of the container is constrained to 20 gallons, if the pressure in the chamber increases from 760 mm Hg to 950 mm Hg, then it can be assumed that the temperature would:

   Are you looking for a data point or a trend? _____

2. The scientist in Study 1 brings the chamber to a temperature of 450°F. Assuming the relationship between temperature and volume is consistent, to which of the following will the volume be closest?

   Are you looking for a data point or a trend? _____

Keyword Clues

There are several additional keywords that will give you more information about "What you know" or "What you need" to know. Becoming familiar with these vocabulary words will help you to recognize keywords, trends, and data points more quickly.

- **Directly proportional** describes two trends that both increase or decrease together.
  Air pollution is <u>directly proportional</u> to the number of cars...

- **Inversely proportional** describes two trends in which one increases while the other decreases.
  The number of mice is <u>inversely proportional</u> to the number of snakes...

- **Ratio** compares one data point to another.
  The <u>ratio</u> of female to male tortoises was 3 : 2...

- **Consistent** implies that a trend follows the same pattern shown in the figure.
  Is the information below <u>consistent</u> with the findings of Scientist 1?

- **Most likely** indicates that you will be predicting a value beyond the range of the graph.
  If the algae doubled, the amount of oxygen would <u>most likely</u>...

- **Expect** sometimes indicates that you will be predicting a value beyond the range of the graph.
  If the temperature dropped to 20°C, the frogs would be <u>expected</u> to...

- **Respectively** indicates that items are listed in the order they are described.
  The first and second fossils were 5 and 2 million years old, <u>respectively</u>...

keep in mind

If two things are "proportional," they are always linked. Remember that "inverse" means "opposite," so if one variable goes up, the other goes down.

Try It Out!

Underline the "What you know" and "What you need" keywords in the questions below.

1. The Stefan-Boltzman relationship states that the luminosity of a star is directly proportional to its area. Using the data in Figure 1, we can expect that the largest star within 5 parsecs of the Sun is:

2. Astronomers have now detected several stars within 10 parsecs of the sun that appear to be orbited by planets. These include Gliese 876 and AU Microscopii, which have predicted luminosities of 0.0124 Solar and 0.1 Solar, respectively. In comparison to the sun, these stars are most likely:

Guided Practice

The 3-Step Method for Data Representation

STEP 1: Highlight keywords.

- *Read the introduction to determine the general focus of the passage.*
- *Read a question and highlight the keywords.*

STEP 2: Use the diagrams.

- *Use the keywords to select the correct headings from the correct diagram.*
- *Find the data point or summarize the trends.*
- *If the information is not in the table, read the text above the table.*

STEP 3: Answer the question.

- *Match the information from the diagram to one of the answer choices.*
- *Check that the text you have underlined and your analysis of the diagram support your answer.*

Use the 3-Step Method for Data Representation to read the passage and answer the questions.

The table below shows some physical properties of common optical materials. The refractive index of a material is a measure of the extent to which light is bent upon entering the material. The transmittance range is the range of wavelengths over which the material is transparent. The chemical resistance of a material is a measure of the material's ability to withstand deterioration or corrosive action from other chemicals.

Material	Refractive index for light of 0.589 μm	Transmittance range (μm)	Useful range for prisms (μm)	Chemical resistance
Lithium fluoride	1.39	0.12–6	2.7–5.5	Poor
Calcium fluoride	1.43	0.12–12	5–9.4	Good
Sodium chloride	1.54	0.3–17	8–16	Poor
Quartz	1.54	0.20–3.3	0.20–2.7	Excellent
Potassium bromide	1.56	0.3–29	15–28	Poor
Flint glass*	1.66	0.35–2.2	0.35–2	Excellent
Cesium iodide	1.79	0.3–70	15–55	Poor

*Flint glass: quartz with lead oxide

1. Which material(s) in Table 1 will transmit light at 25 μm?
   A. Potassium bromide only
   B. Potassium bromide and cesium iodide
   C. Lithium fluoride and cesium iodide
   D. Lithium fluoride and flint glass

2. A student hypothesized that any material with poor chemical resistance would have a transmittance range wider than 10 μm. The properties of which of the following materials contradict this hypothesis?
   F. Lithium fluoride
   G. Flint glass
   H. Cesium iodide
   J. Quartz

Shared Practice

Use the 3-Step Method for Data Representation to read the passage and answer the questions.

Despite their name, blue-green algae are not algae at all, but photosynthesizing bacteria. They are often blue or green in color, but are common in other colors as well, such as pink and red. In large numbers, they appear as a thick slimy coating on aquatic rocks and plants. They are capable of smothering plants, and may release toxins that are harmful to fish. Some varieties of red tide are caused by blue-green algae.

Like other bacteria, blue-green algae reproduce through a process wherein one cell divides completely to form two distinct cells. The average time for a colony of blue-green algae to divide, and its population to double, is called the *generation time*. Table 1 shows the generation times for different varieties of blue-green algae, in the presence of specific nutrients and at specific temperatures.

Because blue-green algae cannot travel in search of necessary nutrients, the population density of a colony also affects its generation time. Figure 1 shows the generation time of a colony of Dactylococcopis salina as a function of its population per square centimeter of water surface at 38°C.

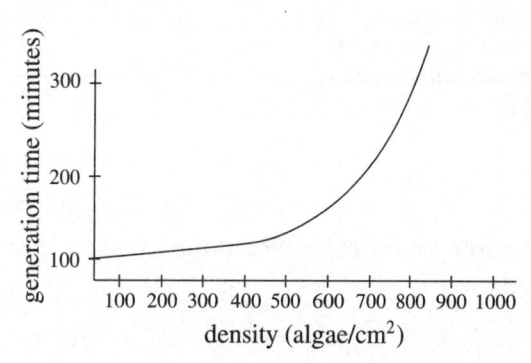

Figure 1

Table 1

Algae	Growth medium	Temperature (°C)	
Aphanothece gelatinosa	Na_3PO_4	30	42
Arthrospira fusiformis	K_4SiO_4	30	103
Chroococcus sonorensis	NH_4NO_3	30	231
Dactylococcopsis salina	Na_3PO_4	38	259
Geitlerinema carotinosum	NH_4NO_3	38	39
Gloecothece rupaestris	K_4SiO_4	38	61
Katagnymene spiralis	Na_3PO_4	38	131
Merismopedia glauca	NH_4NO_3	38	751
Rhabdoderma rubrum	NH_4NO_3	38	225
Synechococcus elongatus	K_4SiO_4	30	321
Synechococcus leopoliensi	Na_3PO_4	30	458
Thermosynechococcus valcanus	K_4SiO_4	38	56

1 On the basis of the data presented in Table 1, if *Arthrospira fusiformis* were placed in a petri dish containing potassium silicate (K_4SiO_4) at 30°C, its generation time would most likely be:

A. less than 50 min.
B. between 51 and 70 min.
C. between 71 and 90 min.
D. between 91 and 110 min.

hint *You do not need to read or understand any of the text to answer this question.*

2. According to Figure 1, if a colony of *Dactylococcopsis salina* covered a pond floor with a density of 400 algae per square centimeter, what would you expect its generation time to be?

   F. About 20 min.
   G. About 50 min.
   H. About 100 min.
   J. About 120 min.

> **hint** Remember to find the keywords that tell you "What you know" and "What you need" to know.

3. Based on the data recorded in Table 1, of the blue-green algae growing in the presence of ammonium nitrate (NH_4NO_3), which of the following took the longest amount of time to double its population?

   A. *Merismopedia glauca*
   B. *Geitlerinema carotinosum*
   C. *Chroococcus sonorensis*
   D. *Rhabdoderma rubrum*

> **hint** Which label in the table means the same thing as the "What you need" keywords: "double its population"?

Name_____ Date_____

4. A tank of water enriched with sodium phosphate (Na$_3$PO$_4$) at 38°C contains an unknown species of blue-green algae. This species was observed to take about 180 minutes to double its population. According to Table 1, it is most likely:

    F. *Geitlerinema carotinosum.*
    G. *Katagnymene spiralis.*
    H. *Rhabdoderma rubrum.*
    J. *Dactylococcopsis salina.*

hint Use Eliminating to check each "What you know" keyword for each possible answer choice.

5. According to the information in Figure 1, what was the approximate population density, in bacteria/cm^2, of the colony of *Dactylococcopsis salina* that was studied in Table 1?

    A. 100
    B. 300
    C. 800
    D. 1,000

hint First, find the generation time of the bacteria in Table 1, then use that generation time to find the population density in Figure 1.

Name_____ Date_____

KAP Wrap

Highlighting keywords to take the ACT isn't like highlighting text for other purposes. Explain how you can decide which words are ACT keywords, and which ones are not.

lesson B
Tables and Figures

Thinking KAP

After two weeks of reading travel guides, browsing the Internet, and talking to people, you and your friends know a lot more about places to go in Mexico. You have decided to make two stops on your journey. First, you will visit the Lacandon Rainforest, one of the most diverse ecosystems in the world. Then, you will scuba dive around Cozumel Island for a few days while you visit your friends' relatives and get a taste of Mexican culture.

The journey from the rainforest to Cozumel is about 500 miles—about the distance from Chicago to Memphis. You do some research into transportation methods, but you still must make a choice.

Transportation	Cost	Time
Plane	$400	2 hours
Train	$200	6 hours
Bus	$100	10 hours

What is the relationship between cost and time? Which method of transportation would you choose?

Strategy Instruction

Using the Figures

Getting information from a figure is done in two parts. To choose the best method of transportation, you first recognized that the decision would be based upon the time and the cost. After you knew what to look for, you found the columns on the table that held that information and looked for trends. Then you used that information to make a decision.

The 3-Step Method for Data Representation uses the same process:

STEP 1: Highlight keywords.

- *Read the introduction to determine the general focus of the passage.*
- *Read a question and highlight the keywords.*

STEP 2: Use the diagrams.

- *Use the keywords to select the correct headings from the correct diagram.*
- *Find the data point or summarize the trends.*
- *If the information is not in the table, read the text above the table.*

STEP 3: Answer the question.

- *Match the information from the diagram to one of the answer choices.*
- *Check that the text you have underlined and your analysis of the diagram support your answer.*

Lesson A was about Step 1: Highlight Keywords. We used this step to find out what information we need to get from a graph and where to find it. In this lesson, we will use those keywords in Step 2: Use the diagrams.

Data Formats on the ACT

There are two main formats that the ACT uses for presenting data: line graphs and tables.

Line Graphs

Line graphs show change over time, or change on a continuous scale:

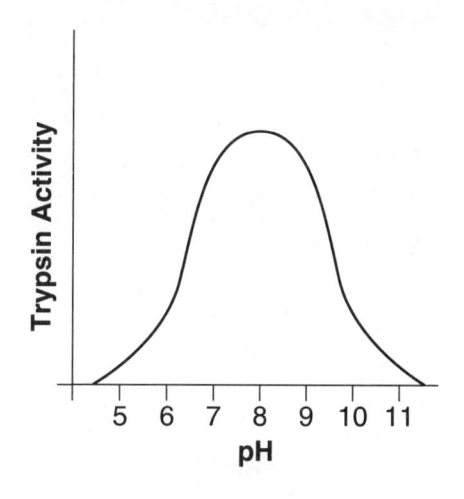

Figure 1

keep in mind

Watch out! There may be a Figure 1 and a Table 1 on the same page for the same experiment.

Tables

Tables may show sequential or non-sequential data. The ACT often will not list data in increasing or decreasing order.

Table 1

Lake	Temperature (°C)	Mercury (ppm)	Fish per 100 m³
1	16.1	1	3.5
2	16.3	67	0
3	15.8	3	5.7
4	16.0	4	9.5

When approaching tables and figures on the ACT, begin with the keywords from a question. First, find the table or figure, then find the column, row, or axis that matches the "What you know" keywords. Find the data point or trend that matches the "What you know" keywords, and put your finger or pencil point on it. Then find the column, row, or axis that matches the "What you need" keywords. Finally, use "What you know" and "What you need" to find the point or trend that answers the question.

Unfamiliar Data Formats on the ACT

The ACT also includes several graphical formats for presenting information that may be unfamiliar. However, you can read any type of graph by using the **axes** and **labels** to follow the same steps for simpler figures.

Scatter Plots

Scatter plots show each data point separately. Sometimes trends are found in clusters, or they may approximate lines.

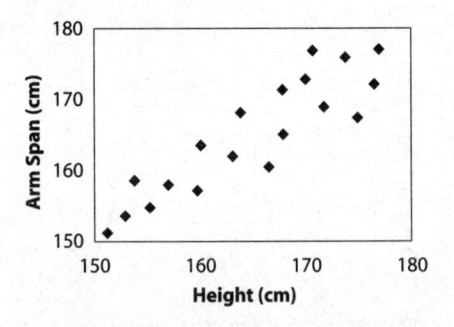

Figure 2

Vertical Line Graphs

Sometimes line graphs that represent depth are presented vertically. These line graphs are just like the line graphs you have seen before.

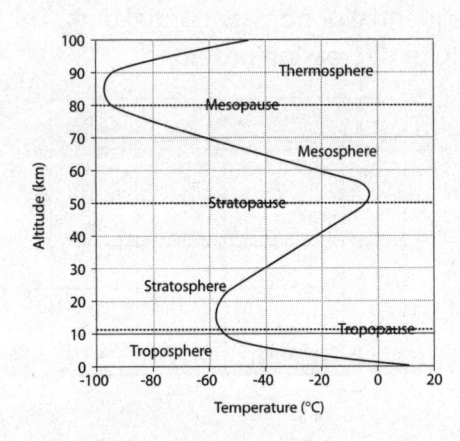

Figure 3

Shaded Tables

Some tables show regions that are shaded or blank instead of showing numbers.

Table 2	Site			
Fossil type	1	2	3	4
Brachiopod	■			■
Cephalopod	■			
Clam		■		■
Land snail		■		
Mammoth tusk		■		
Mastodon tusk		■	■	
Water snail	■			■

Try It Out!

Underline keywords, then use the information above to answer the questions below.

1. According to Table 1, what is the temperature of Lake 2? _____

2. Which of the following statements correctly describes the trend shown in Figure 1? As pH increases, trypsin activity:
   - **A.** increases, then decreases.
   - **B.** decreases, then increases.
   - **C.** increases.
   - **D.** decreases.

3. According to Figure 1, at what pH is trypsin most effective? _____

4. According to Figure 2, height and arm span are:
   - **F.** inversely proportional.
   - **G.** directly proportional.
   - **H.** exactly equal.
   - **J.** not related.

5. According to Table 2, which site(s) contained brachiopods, clams, and water snails, but not cephalopods? _____

6. According to Figure 3, what is the temperature at the stratopause?
   _____

Working with Multiple Axes

Figures and tables on the ACT often show several axes on the same graph. This saves space and allows several pieces of information to be shown together. Remember to find the correct axis and trend line before answering the question.

Line Graphs with Multiple Scales

Line graphs may have units labeled on two *y*-axes or two *x*-axes. For these figures, it is necessary to use the key to determine which trend to look at. Then use the axes to determine which scale to use.

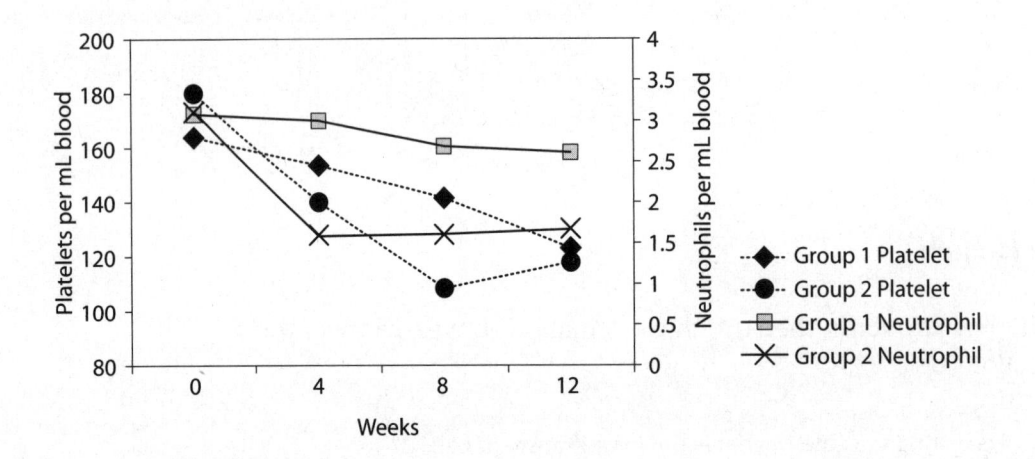

Figure 4

Tables with Multiple Sets of Data

Tables may also mix many types of information. In the table below, average mass and vitamin C are independent values, while the caloric ratio is divided into three sub-sections: carbohydrates, fats, and protein. Similarly, there are two families of fruit represented: Rutaceae include oranges, lemons, and grapefruit, while Rosaceae include apples, Asian pears, and pears.

	Table 3					
Fruit		Average mass (g)	Vitamin C (% DV)	Caloric ratio (%)		
				Carbohydrates	Fats	Protein
Rutaceae	Orange	141	106	91	4	5
	Lemon	84	74	78	9	13
	Grapefruit	256	146	90	3	7
Rosaceae	Apple	138	11	95	3	2
	Asian pear	122	8	91	5	4
	Pear	166	12	96	2	2

Try It Out!

Underline keywords; then use the information in Table 3 to answer the questions below.

1. According to Table 3, which fruit has the highest level of protein, by percent calories? _____

2. In what category (average mass, vitamin C, or caloric ratio) is there a significant difference between Rutaceae and Rosaceae? _____

3. According to Figure 4, what is the average neutrophil count for members of group 2 during week 4? _____

4. High platelet and neutrophil counts are necessary for maintaining good health. Which group is in a healthier condition, in terms of platelets and neutrophils, at the end of the experiment described in Figure 4? _____

Guided Practice

The 3-Step Method for Data Representation

STEP 1: Highlight keywords.

- Read the introduction to determine the general focus of the passage.
- Read a question and highlight the keywords.

STEP 2: Use the diagrams.

- Use the keywords to select the correct headings from the correct diagram.
- Find the data point or summarize the trends.
- If the information is not in the table, read the text above the table.

STEP 3: Answer the question.

- Match the information from the diagram to one of the answer choices.
- Check that the text you have underlined and your analysis of the diagram support your answer.

Use the 3-Step Method for Data Representation to answer the questions.

Passage I

pH is a quantitative measure of hydrogen ion concentration. A solution with a pH between 0 and 7 is acidic, while a solution with a pH between 7 and 14 is basic. An acid-base indicator is a weak acid or base that is sensitive to the hydrogen ion concentration of a solution and changes color at a known pH. At pH levels above or below the indicator range, the indicator remains the same color.

The figure below shows various acid-base indicators, and the pH ranges at which they exhibit different colors.

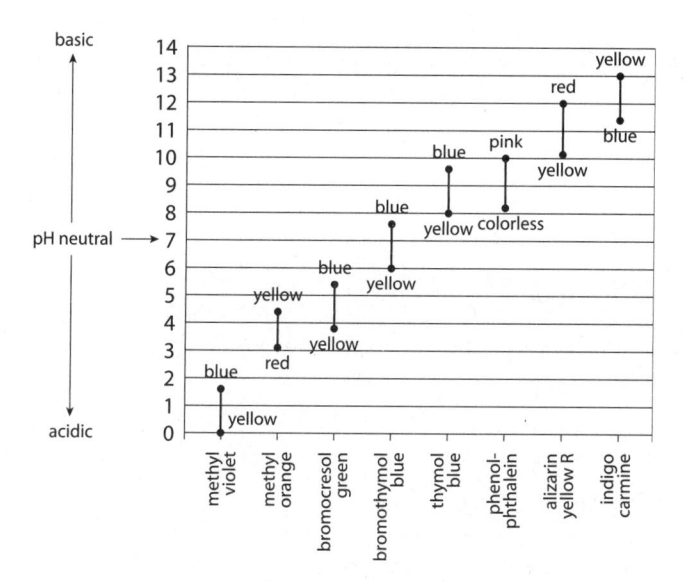

2. A scientist uses alizarin yellow R to measure the pH of a solution and finds that it becomes red. The scientist hypothesizes that the pH of the solution is 12. Is this hypothesis correct?

F. Yes, because alizarin is yellow above a pH of 12.
G. Yes, because alizarin is red above a pH of 12.
H. No, because alizarin is orange above a pH of 12.
J. No, because alizarin does not indicate a precise pH above 11.

1. A chemist is running an experiment in a solution that becomes basic when the reaction is complete. According to the figure, the reaction is complete when it causes:

A. bromothymol blue to become blue.
B. all indicators to become violet.
C. a white solid to appear.
D. bromocresol green to become yellow.

Shared Practice

Use the 3-Step Method for Data Representation to read the passage and answer the questions.

Passage I

The extent to which a solute will dissolve in a given solvent is dependent on several factors, including the temperature and pressure conditions and the electrochemical natures of the solute and solvent. A high school chemistry teacher assigned 24 students the project of measuring the solubilities in distilled water of several pairs of common sodium (Na) and potassium (K) salts at various temperatures. The pairs were NaCl and KCl, $NaNO_3$ and KNO_3, and $NaClO_3$, and $KClO_3$. All measurements were conducted under normal atmospheric pressure. After pooling and averaging all the data, the students plotted solubility curves to produce the graph below.

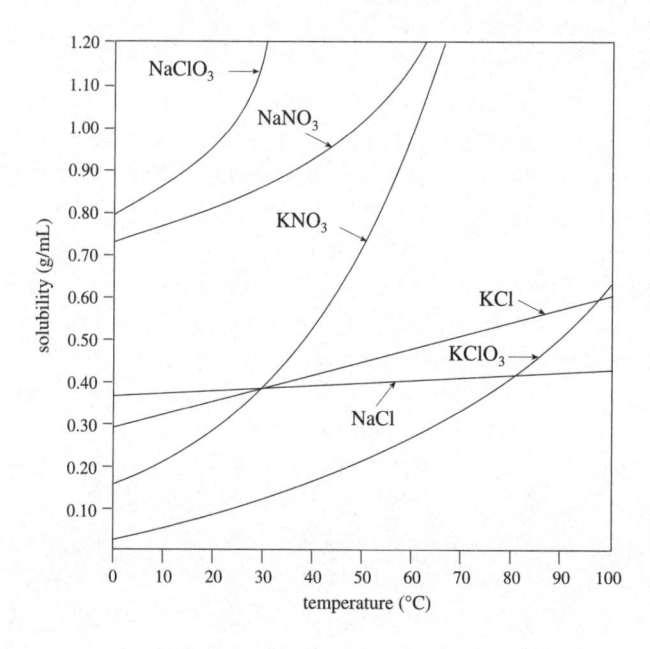

1. For the salts given, which of the following conclusions can be drawn concerning the relationship between solubility and temperature?
   A. As the temperature increases, the solubility increases.
   B. As the temperature increases, the solubility decreases.
   C. As the temperature increases, the solubility remains the same.
   D. Solubility and temperature are unrelated.

hint *Use the "What you know" and "What you need" keywords to find the trends of one salt. Then make sure the others match.*

Name_____   Date_____

2. How does the solubility data for the chloride salts (NaCl and KCl) differ from the data for the nitrate (NaNO$_3$ and KNO$_3$) and chlorate (NaClO$_3$ and KClO$_3$) salts?

F. The chloride salts have a lower solubility than the other salts at all temperatures.
G. The chloride salts have a higher solubility than the other salts at all temperatures.
H. The solubility of the chloride salts increases exponentially, while the other salts increase linearly.
J. The solubility of the chloride salts increases linearly, while the other salts increase exponentially.

hint Circle the labels for the chloride salts on the graph to help you find them while you eliminate wrong answers.

3. At 65°C, which of the following salts have solubilities over 0.65 g/mL?

A. All sodium (Na) salts
B. All potassium (K) salts
C. Both nitrate (NO$_3$) salts
D. Both chlorate (ClO$_3$) salts

hint Use the edge of a piece of paper to draw a vertical line at 65°C and a horizontal line at 0.65 g/mL. Then look for salts that intersect the vertical line above the horizontal line.

B

74 ACT ADVANTAGE SCIENCE   © 2006 Kaplan, Inc.

4. A student wants to make a 1-liter solution containing equal amounts of KNO_3 and $KClO_3$ in distilled water at 50°C. To do this, the student should use:

F. equal amounts of KNO_3 and $KClO_3$.
G. more KNO_3 than $KClO_3$.
H. more $KClO_3$ than KNO_3.
J. KCl to balance the KNO_3 and $KClO_3$.

hint *If more of the salt dissolves, do you need to add more or less distilled water?*

5. Based on the data in the passage, which of the following solutes exhibits the greatest variation of solubility with temperatures between 0°C and 60°C?

A. NaCl
B. KCl
C. KNO_3
D. $KClO_3$

hint *The variation is the difference between the highest and the lowest value.*

Name_____ Date_____

KAP Wrap

The first two steps of the 3-Step Method for Data Representation are written below. After each step, write one piece of advice you would give to someone trying to use the method for the first time.

STEP 1: Highlight keywords.

- **Read the introduction to determine the general focus of the passage.**

Advice: _____

- **Read a question and highlight the keywords.**

Advice: _____

STEP 2: Use the diagrams.

- **Use the keywords to select the correct headings from the correct diagram.**

Advice: _____

- **Find the data point or summarize the trends.**

Advice: _____

- **If the information is not in the table, read the text above the table.**

Advice: _____

UNIT 2: DATA REPRESENTATION PASSAGES
LESSON B: TABLES AND FIGURES

lesson C: Using Patterns

ReKAP

Review the strategies from Lessons A and B. Then fill in the blanks with what you have learned.

1. When looking for keywords, first find the ones that tell you "What you _____," such as axes, _____, and _____. Then look for "What you _____" keywords, which tell you where to find the information and if it is a _____ or a _____.

2. When approaching a new passage, you should always read the _____, but you do not need to read the rest unless it is necessary to answer a question.

3. If the "What you know" keywords contain a data point, and the "What you need" keywords do not state what you are looking for, you can assume that _____.

4. List two things that you should NOT underline when looking for "What you know" or "What you need" keywords:

   _____

   _____

Strategy Instruction

Making Predictions

Some ACT questions will ask you to predict data based upon a trend. This may involve extending the trend beyond the highest or lowest data point available, or predicting a new point between two other points. If there are no lines in the area of the graph for which you are predicting data, use the edge of a sheet of paper to draw a straight line to connect the points and extend them beyond the edges of the graph.

Try It Out!

Underline the keywords, then use the figure below to answer the questions. Remember to look for the keywords *expect* and *most likely*, which may indicate that you will be predicting the data point.

Crickets are cold-blooded organisms whose metabolic rate changes with the temperature. The diagram below gives the average number of cricket chirps per 15-second interval at different temperatures.

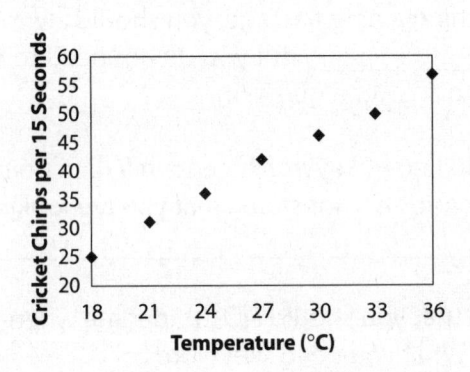

1. As the temperature increases to 41°C, crickets could be expected to chirp:
   A. 50 times per 15 seconds.
   B. 62 times per 15 seconds.
   C. 67 times per 15 seconds.
   D. 80 times per 15 seconds.

2. In a separate study, crickets were observed to chirp 38 times per 15 seconds. Using the data in the figure, the temperature is most likely:
   F. 24°C.
   G. 26°C.
   H. 27°C.
   J. 64°C.

Examining New Data

Some questions may ask you to compare a new data point to the data in a figure or table. To answer these questions, first draw the new data point on the existing figure. Then use the keywords to make comparisons.

Try It Out!

Read the passage and examine the figure. Then answer the questions that follow.

Tensile strength is a measure of the pull stress (in force per unit area or MPa) required to break a given specimen. The tensile strength of four polymers was tested at a range of temperatures, as shown in the figure below.

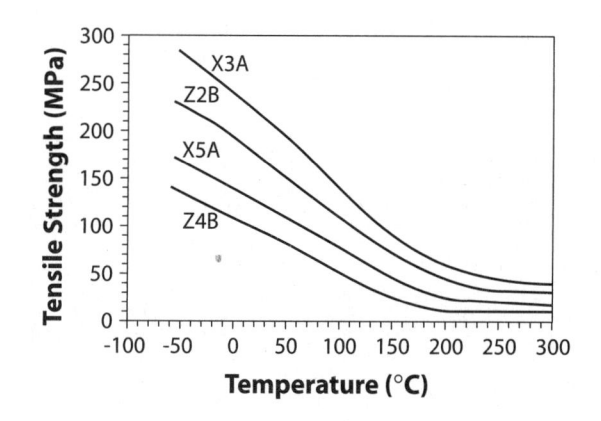

1. A company develops a new polymer, Polymer A. At 0°C, Polymer A has a tensile strength of 130 MPa. Based on the results of the experiment above, which of the following correctly lists the tensile strength of the five polymers at 0°C from *lowest* to *highest*?
   A. X3A, Z2B, X5A, Polymer A, Z4B
   B. X3A, Z2B, Polymer A, X5A, Z4B
   C. Z4B, X5A, Polymer A, Z2B, X3A
   D. Z4B, Polymer A, X5A, Z2B, X3A

2. Suppose a company produces a new polymer, Polymer B, which may be able to replace Polymer Z2B in a machine that contains boiling water. At 100°C, Polymer B has a tensile strength of 155 MPa. In comparison to Z2B, Polymer B has a tensile strength which is:
   F. 55 MPa higher.
   G. 55 MPa lower.
   H. 15 MPa higher.
   J. 15 MPa lower.

Name_____ Date_____

Test Practice Unit 2

When your teacher tells you, carefully tear out this page. Then begin working.

1. Ⓐ Ⓑ Ⓒ Ⓓ   15. Ⓐ Ⓑ Ⓒ Ⓓ

2. Ⓕ Ⓖ Ⓗ Ⓙ   16. Ⓕ Ⓖ Ⓗ Ⓙ

3. Ⓐ Ⓑ Ⓒ Ⓓ

4. Ⓕ Ⓖ Ⓗ Ⓙ

5. Ⓐ Ⓑ Ⓒ Ⓓ

6. Ⓕ Ⓖ Ⓗ Ⓙ

7. Ⓐ Ⓑ Ⓒ Ⓓ

8. Ⓕ Ⓖ Ⓗ Ⓙ

9. Ⓐ Ⓑ Ⓒ Ⓓ

10. Ⓕ Ⓖ Ⓗ Ⓙ

11. Ⓐ Ⓑ Ⓒ Ⓓ

12. Ⓕ Ⓖ Ⓗ Ⓙ

13. Ⓐ Ⓑ Ⓒ Ⓓ

14. Ⓕ Ⓖ Ⓗ Ⓙ

UNIT 2: DATA REPRESENTATION PASSAGES
LESSON C: USING PATTERNS

SCIENCE TEST
15 Minutes—16 Questions

DIRECTIONS: There are three passages in this test. Each passage is followed by several questions. After reading a passage, choose the best answer to each question and fill in the corresponding oval on your answer sheet. You may refer to the passages as often as necessary.

You are NOT permitted to use a calculator on this test.

Passage I

Samples of sap were collected from 4 different species of maple tree one year in April and again in November. Each sample was then analyzed. Tables 1 and 2 show the volume, color, specific gravity, and concentration of glucose (sugar) in the April and November samples, respectively. Specific gravity is calculated using density, or weight per unit volume, in the following equation:

$$\text{specific gravity} = \frac{\text{density of sample}}{\text{density of water}}$$

The normal range for the specific gravity of tree sap is 2.023–4.527.

Table 1
Sap samples collected April 10

Species	Volume (mL)	Color*	Specific gravity	Glucose (g/L)
A	360	10	4.501	102.3
B	418	2	3.215	76.4
C	432	7	3.722	87.2
D	476	3	3.198	75.2

*Note: Color was assigned using the following scale: 0 = clear; 10 = very dark.

Table 2
Sap samples collected November 18

Species	Volume (mL)	Color*	Specific gravity	Glucose (g/L)
A	100	3	2.835	20.1
B	124	1	2.523	18.6
C	145	2	2.792	19.5
D	173	1	2.499	18.5

*Note: Color was assigned using the following scale: 0 = clear; 10 = very dark.

1. Based on the information presented, which of the following samples most likely has the highest water content per milliliter?
   A. The April sample from species A
   B. The April sample from species D
   C. The November sample from species A
   D. The November sample from species D

2. Does the data in Tables 1 and 2 support the conclusion that as sap sample volume increases, sap color darkens?
   F. Yes, because sap samples with the greatest volumes had the greatest color values.
   G. Yes, because sap samples with the greatest volumes had the lowest color values.
   H. No, because sap samples with the greatest volumes had the lowest color values.
   J. No, because there is no relationship between sap sample volume and color value.

3. Based on the information provided, as the concentration of glucose in tree sap rises, the specific gravity of the sap:
   A. increases only.
   B. decreases only.
   C. increases, then decreases.
   D. decreases, then increases.

GO ON TO THE NEXT PAGE.

4. When selecting sap to make maple syrup, trees with the darkest, sweetest, and thickest syrup are preferred. Based on the information provided, which species would be **best** for making syrup?
   F. Species A
   G. Species B
   H. Species C
   J. Species D

5. From which of the following samples would one milliliter of sap weigh the least?
   A. The April sample from species A
   B. The April sample from species B
   C. The November sample from species C
   D. The November sample from species D

GO ON TO THE NEXT PAGE.

Passage II

Scientists are interested in the history of global climate change to determine if dramatic global changes in temperature are normal, or if global warming is a novel phenomenon caused by human activity. Although scientists have monitored surface water temperature for the last hundred years, they must be able to determine how temperature has changed over thousands of years to understand this complex phenomenon.

Experiment 1

Scientists can estimate the temperature of the ocean water surfaces from hundreds of thousands of years ago by drilling through ice samples to test the oxygen content of water from different eras. The deeper the sample, the older it is. The partial pressure of oxygen in the frozen sample is directly related to seawater temperature, and a one part per million decrease in the $\delta^{18}O$ measurement is calculated to be related to an approximate 1.5°C decrease in temperature at the time the water evaporated from the ocean. Because of the complexity of these calculations, scientists calibrate their data by looking at similar time period samples from two different sites, as shown in Figure 1.

Experiment 2

It has been hypothesized that human-based increases in CO_2 have contributed to recent climate changes. A team of scientists is investigating the relationship between carbon dioxide (CO_2) in the atmosphere and atmospheric temperature, using the measured global CO_2 levels and average temperatures for the last 150 years, as shown in Figure 2.

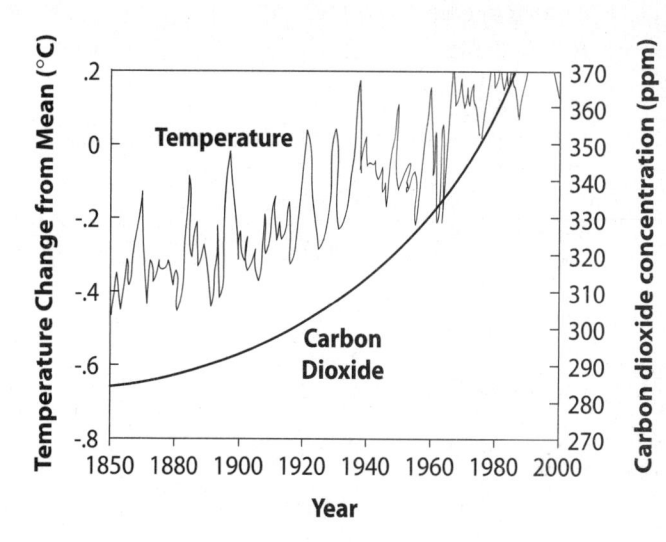

Figure 2

6. Humans have been polluting the air with commercial chemicals for over 100 years. Which figure better supports the hypothesis that recent human industry is a cause of global warming?

   F. Figure 1, because the temperature has only varied by 12°C over the past 120 thousand years.
   G. Figure 1, because trends towards modern temperatures have been seen in the past.
   H. Figure 2, because the mean temperature has steadily increased over the last century.
   J. Figure 2, because from 1850-1950 it was cooler than the mean.

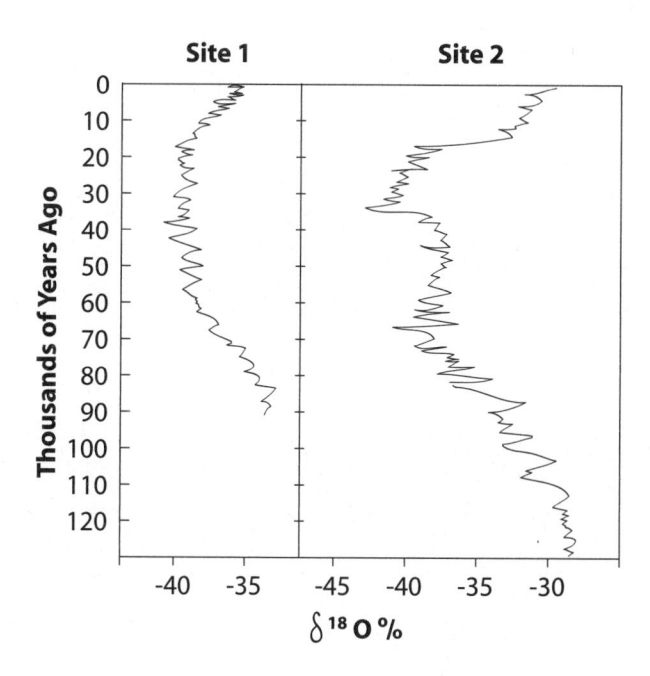

Figure 1

7. According to the data from both Site 1 and Site 2 in Figure 1, at what point was the global temperature lowest?
   A. 0–10 thousand years ago
   B. 30–40 thousand years ago
   C. 70–80 thousand years ago
   D. 100–110 thousand years ago

8. Based upon the data in Figure 2, in the year 2020, the temperature change from the mean is expected to be:
   F. −0.2°C.
   G. 0.2°C.
   H. 0.4°C.
   J. The trends are not clear enough to use for predictions.

9. Scientists are interested in situations that cause patterns to change dramatically. According to the data in Figure 1, which period would *best* lead scientists to a better understanding of causes of global climate changes?
   A. 0–10 thousand years ago
   B. 10–20 thousand years ago
   C. 20–30 thousand years ago
   D. 30–40 thousand years ago

10. In the 20 years following 1960, CO_2 levels rose approximately:
    F. 10 ppm.
    G. 20 ppm.
    H. 30 ppm.
    J. 40 ppm.

GO ON TO THE NEXT PAGE.

Passage III

Menstruation begins in women at an average age of 13.9 years and ceases in a process called menopause at an average age of 51, though individuals may vary widely from these dates. The function of the menstrual cycle is to prepare a single egg for maturation, ovulation, and fertilization every month. Some women may have menstrual periods as frequently as every two weeks, as infrequently as every 3 months, or even once a year during perimenopause, the time surrounding the onset of menopause. However, the majority of women during most of their lives menstruate approximately once every 28 days, a cycle which may have been attuned to the new moon prior to artificial lighting.

The menstrual cycle is controlled by the interactions of four major hormones: follicle stimulating hormone (FSH), estrogen, luteinizing hormone (LH), and progesterone. Their effects on the follicle which secretes the egg and the endometrial lining of the uterus are shown below.

Ovarian and Menstrual Cycles

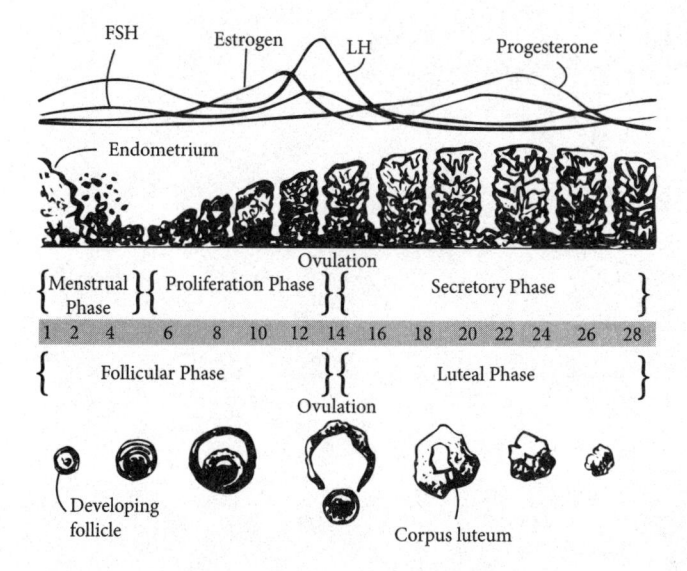

11. The best predictor of ovulation is a high level of:
    A. follicle stimulating hormone (FSH).
    B. estrogen.
    C. luteinizing hormone (LH).
    D. progesterone.

12. During the menstrual phase, the blood stored in the endometrium is released, along with an unfertilized egg from the uterus. At the same time that the menstrual phase occurs, what occurs in the ovaries, where follicles are stored?
    F. A new follicle initiates development.
    G. The follicle releases an egg.
    H. The egg within the follicle becomes larger.
    J. The follicle becomes a corpus luteum.

13. The follicular phase is characterized by a rise in LH and FSH. On which day of the average menstrual cycle do these increases begin?
    A. 1
    B. 9
    C. 24
    D. 28

14. Perimenopause is often characterized by estrogen dominance, extremely high levels of estrogen and low levels of progesterone. Thus, instead of completing regular menstrual cycles, women remain for long periods of time in:
    F. the menstrual phase.
    G. the proliferation phase.
    H. ovulation.
    J. the secretory phase.

15. During the luteal phase, women's bodily temperatures increase, which provides incubation in case the egg has been fertilized. When it is clear that fertilization has not occurred, the body temperature decreases for the remainder of the cycle. Which hormone is responsible for this increase in body temperature?
    A. Follicle stimulating hormone (FSH)
    B. Estrogen
    C. Luteinizing hormone (LH)
    D. Progesterone

16. Which two hormones follow the same patterns of increasing and decreasing throughout the menstrual cycle?
    F. LH and FSH
    G. Progesterone and LH
    H. Estrogen and FSH
    J. Estrogen and progesterone

END OF TEST.

STOP! DO NOT TURN THE PAGE UNTIL TOLD TO DO SO.

Name_____ Date_____

KAP Wrap

Suppose you meet a man or woman from another country who does not speak the same language as you. He or she has just moved to your city and you feel like welcoming him or her in some way. How could you describe your city to this person, other than through spoken language? What would be the best way to teach the person about your city? How could you communicate your likes and dislikes to this person?

Unit 3
Data Representation Passages Part 2

lesson A
Understanding Experiments

Thinking KAP

Your tour guide leads you and your friends into the Lacandon Rainforest. There are brightly colored birds, towering trees, and chattering monkeys. When your group stops in a clearing, you notice a huge aloe plant. It looks a lot like the tiny one you keep in a pot at home, but this one has yellow flowers and it's nearly as tall as you are.

One friend thinks this is a different species of aloe than you have at home, which is why it is so big. Your other friend thinks that the plants are the same species, but that different growing conditions made the plants look different. What experiment could you perform to figure out who is correct?

Strategy Instruction

Understanding Experiments

Although most questions on the ACT involve finding or interpreting data, there may be a few questions about the procedure of an experiment. In general, these questions will ask you to predict what will happen when experimental conditions are changed.

Review the elements of an experiment below.

keep in mind

The experiments found in the ACT vary by subject—but because all scientists follow the same approach, focus on the scientific method rather than specific subject-based details.

Analyzing the Experiment

- **Independent Variable**
  The independent variable is what the experimenter wants to know more about.
  The ACT may refer to this as the "factor that is varied" in the experiment.

- **Dependent Variable**
  The dependent variable is what is measured.
  The ACT may refer to this as the "factor that is NOT directly controlled."

- **Control**
  The control is a trial or group that is "normal." It is used to calibrate the equipment or show that the independent variable had an effect.
  The ACT may identify the control and ask why it is present in the experiment.

Try It Out!

Analyze the elements of the experiment below.

Ginger has been used for 2,500 years in China and India to prevent headaches, nausea, rheumatism, and colds. A group of scientists is comparing the effects of powdered ginger to over-the-counter motion sickness medications for preventing nausea.

Experiment 1

A team of doctors asked for volunteers among families with children who reported problems with motion sickness, and who drove for over 1 hour continuously at least five days a week. In all, 230 children and 192 adults participated in the full study. All family members were asked to take a tablet one hour before riding in the car, then record the number of times that any family member experienced nausea or vomited while in the car or within 30 minutes afterward.

During the first month, families were given a placebo (a tablet containing starch instead of medication). During the second, they were given a tablet containing powdered ginger; during the third, a non-prescription motion sickness medication; and during the fourth, a prescription motion sickness medication. The participants, and the nurses dispensing the tablets, were not told the contents of any of the tablets. The results are shown below in Table 1.

Table 1				
Month	Tablet	Incidence reports of children		Incidence reports of adults	
		Motion sickness	Vomiting	Motion sickness	Vomiting
1	Placebo	554	35	387	2
2	Ground ginger	87	3	15	0
3	Non-prescription medication	190	8	71	1
4	Prescription medication	122	15	55	2

1. In Experiment 1, the factor that is varied is:
   A. the number of children.
   B. the contents of the tablets.
   C. the time spent driving.
   D. motion sickness.

2. Which factors are NOT directly controlled in Experiment 1?
   F. Vomiting and nausea
   G. Children and adults
   H. Ginger and motion sickness medication
   J. Days driving and months

3. Why were the participants given a placebo tablet for the first month?
   A. It increased motion sickness during the first month, allowing it to be studied later.
   B. It allowed doctors to determine how effective ginger and the motion sickness medication are in comparison to no treatment.
   C. Doctors believe the placebo might be helpful in preventing motion sickness.
   D. Doctors used the placebo to determine the correct dosage of ginger or motion sickness medication for children and adults.

Comparing New Experiments

A scientist must always look for new evidence. Part of the scientific method involves using the results of one experiment to design a new experiment or examining how the results of a new experiment influence existing theories.

The ACT may ask you to identify which areas have not been covered by an experiment, or apply a new formula or idea to the results. Treat these questions like other data representation questions. Highlight "what you know" and "what you need" keywords, then look for them in the data. If either category does not appear, the answer may be "there is not enough information."

Try It Out!

Underline relevant keywords and answer the following two questions about the previous experiment.

4. A patient is interested in using ginger to treat migraine headaches. Based upon the results of Experiment 1, it is likely that:
   F. ginger will decrease the intensity and frequency of migraine headaches.
   G. ginger will increase the intensity and frequency of migraine headaches.
   H. ginger will decrease migraines if a stomach ache is present at the same time.
   J. The results of Experiment 1 are not sufficient to draw a conclusion.

5. The *success rate* of a treatment is the ratio of successful treatments to total patients treated. Which treatment has the highest success rate?
   A. Placebo
   B. Ground ginger
   C. Non-prescription medication
   D. Prescription medication

Experimental Design

Questions about the design of an experiment are generally "Big Idea" questions that require you to synthesize information from the entire passage. These questions may ask you to do any of the following:

- explain how changing the design would change the results.
- explain how changing one part of the design would change another part of the design.
- explain why parts of a design were performed.
- identify the figure that illustrates the experimental design.
- identify problems in a design that could cause errors in the results.
- identify a new experiment that would give more information about the same ideas.

The good news about design questions is that they are rare. When you find an experimental design question, answer all of the other questions about the passage first. Then examine each of the answer choices and reread the passage to see which one makes the most sense. If you're running out of time, make a quick educated guess and move on. Experimental design questions can eat up your time, but you still only get one point.

Try It Out!

Answer the following two "experimental design" questions about the previous experiment.

6. If the experiment were changed to include only adults, how would the results of the study change?
   F. There would be fewer reports per participant of both vomiting and motion sickness.
   G. There would be more reports per participant of both vomiting and motion sickness.
   H. There would be more reports per participant of vomiting but fewer reports of motion sickness.
   J. There would be more reports per participant of motion sickness but fewer reports of vomiting.

7. One researcher notices that the number of people vomiting increases continuously from month 2 to month 4 in both children and adults. The researcher hypothesizes that this might occur because the participants stop driving carefully as their symptoms decrease, rather than because the prescription medication is less effective. If the experiment were repeated, how could it be modified to address this hypothesis?
   A. Have participants take all four tablets at the same time.
   B. Begin the study with the prescription medication.
   C. Remove the prescription medication from the study.
   D. Have participants take twice as much of the prescription medication.

Guided Practice

The 3-Step Method for Data Representation

☝ **STEP 1: Highlight keywords.**

- *Read the introduction to determine the general focus of the passage.*
- *Read a question and highlight the keywords.*

✌ **STEP 2: Use the diagrams.**

- *Use the keywords to select the correct headings from the correct diagram.*
- *Find the data point or summarize the trends.*
- *If the information is not in the table, read the text above the table.*

🤟 **STEP 3: Answer the question.**

- *Match the information from the diagram to one of the answer choices.*
- *Check that the text you have underlined and your analysis of the diagram support your answer.*

Use the 3-Step Method for Data Representation to read the passage and answer the questions.

Ground-level ozone gas is a major component of urban smog. It is not emitted directly as a pollutant, but is formed through a complex set of chemical reactions involving hydrocarbons, nitrogen oxides, and sunlight. The rate of ozone production increases with sunlight intensity and temperature, and therefore peaks during hot summer afternoons.

The hydrocarbons and nitrogen oxides from which ozone is formed come primarily from fossil-fuel-burning engines. Students at an urban high school hypothesized that their school buses emit more hydrocarbons when the engines idle for three minutes than they do by turning off their engines and then restarting them. The students performed a series of experiments to test their hypothesis.

Experiment 1

Students connected a collection bag to the exhaust pipe of one of their school buses. The bus's engine was started, and the exhaust was captured by the collection bag. A syringe was then used to extract a 5-mL sample of the exhaust. The exhaust was injected into a gas chromatograph, which separates mixtures of gases into their individual components. Students compared the exhaust to mixtures of known hydrocarbon concentration samples to determine what percent, by volume, of the sample was composed of hydrocarbons. The bus was started, and samples of the exhaust were collected and extracted at 30-second intervals. Exhaust samples from two other buses were also collected. The results are reproduced below in Table 1.

Table 1			
Time after starting (sec)	Percent of hydrocarbons in the exhaust		
	Bus 1 1984 Model X	Bus 2 1982 Model X	Bus 3 1996 Model X
30	10.3	9.0	5.4
60	11.2	9.8	4.9
90	12.0	13.2	6.0
120	10.5	22.9	4.9
150	9.6	21.0	4.5
180	9.5	20.1	4.2
210	9.4	19.2	4.2
240	9.3	19.2	3.8

Experiment 2

The exhaust of the buses was collected and tested again after the buses had been allowed to idle for 15 minutes, using the same procedure as Experiment 1. The results are reproduced below in Table 2.

Table 2			
Time after starting (min)	Percent of hydrocarbons in the exhaust		
	Bus 1 1984 Model X	Bus 2 1982 Model X	Bus 3 1996 Model X
15.0	5.5	6.1	1.8
15.5	5.5	6.1	1.8
16.0	5.4	6.2	1.8
16.5	5.4	6.2	1.8
17.0	5.4	6.1	1.8
17.5	5.4	6.1	1.8
18.0	5.3	6.1	1.8
18.5	5.3	6.1	1.8

1. In order to pass annual inspection, many states require that buses undergo emissions testing. If the goal of the test is to determine the peak emissions, then when, based on the results of the students' experiments, should the exhaust be sampled?
   A. After 30–60 seconds
   B. After 90–120 seconds
   C. After 210–240 seconds
   D. After 15 minutes

2. The main purpose of Experiment 2 was to:
   F. determine the percentage of hydrocarbons in the exhaust of a warm, idling bus.
   G. calibrate the gas chromatograph.
   H. determine the percentage of nitrogen oxide in a bus's exhaust.
   J. test the effectiveness of the exhaust collection bag.

3. Which of the following best explains why the students collected exhaust samples in a collection bag?
   A. To keep air and other gases from contaminating the exhaust samples
   B. To allow outside air and gases to mix with the exhaust samples equally
   C. To capture only the hydrocarbons in the exhaust
   D. To filter out sediments from the exhaust

4. Which factor of Experiment 1 did the students vary?
   F. The number of 30-second intervals recorded
   G. The instruments used to collect the exhaust samples
   H. The age of the buses tested
   J. The amount of exhaust collected from the buses

Shared Practice

Use the 3-Step Method for Data Representation to read the passage and answer the questions.

Passage I

Osmosis is the movement of water from an area with a lower concentration of dissolved material to an area with a higher concentration of dissolved material. The result of osmosis is equilibrium, the same concentration of dissolved material everywhere. To prevent osmosis, external pressure may be applied to the area with the higher concentration of dissolved material. *Osmotic pressure* is the external pressure required to prevent osmosis.

An apparatus is used to test the osmotic pressure of various solvents. A semi-permeable membrane, which allows solvents to pass through, but not large molecules such as sucrose, is placed in the middle of a container. A solution of the solvent and a large molecule such as sucrose is placed on one side of the apparatus, and a pure solvent is placed on the other. A piston is placed above the solution, and the pressure necessary to keep the piston from moving is recorded. A dye may also be placed in the pure solvent.

Experiment 1

The osmotic pressure of water was tested using four different concentrations of sucrose at two different temperatures. The results are shown in Table 1.

Table 1		
Concentration of sucrose solution (mol/L)	Temperature (K)	Osmotic pressure (atm)
1.00	298.0	24.47
0.50	298.0	12.23
0.10	298.0	2.45
0.05	298.0	1.22
1.00	348.0	28.57
0.50	348.0	14.29
0.10	348.0	2.86
0.05	348.0	1.43

Experiment 2

Sucrose solutions of four different organic solvents were investigated with two different concentrations of sucrose. All trials were held at 298 K. The results are shown in Table 2.

Table 2		
Solvent	Concentration of sucrose solution (mol/L)	Osmotic pressure (atm)
Ethanol	0.50	12.23
Ethanol	0.10	2.45
Acetone	0.50	12.23
Acetone	0.10	2.45
Diethyl ether	0.50	12.23
Diethyl ether	0.10	2.45
Methanol	0.50	12.23
Methanol	0.10	2.45

Name_____ Date_____

1. In Experiment 1, the scientists investigated the effect of:
   A. solvent and concentration on osmotic pressure.
   B. volume and temperature on osmotic pressure.
   C. concentration and temperature on osmotic pressure.
   D. temperature on atmospheric pressure.

   hint *Answer all the questions about Experiment 1 first.*

2. According to the passage, a dye is sometimes placed in the pure solvent when osmotic pressure is tested. What is the *most likely* purpose of this?
   F. The dye shows when osmosis is completed.
   G. The dye shows the presence of ions in the solutions.
   H. The dye is used to show which side is the pure solvent.
   J. The dye is used to show how much water passes through the membrane.

   hint *This is a hard question, so save it for last. Find where "dye" is mentioned in the passage, reread the passage once, eliminate wrong answers, and guess.*

3. According to the experimental results, osmotic pressure is dependent upon the:
   A. solvent and temperature only.
   B. solvent and concentration only.
   C. temperature and concentration only.
   D. solvent, temperature, and concentration.

   hint *Look for differences in the data based upon each keyword in the answers.*

4. A 0.10 mol/L aqueous sucrose solution is separated from an equal volume of pure water by a semipermeable membrane in an open container. What will the condition of the liquids be when equilibrium is reached?
   F. The pure water side will have more volume.
   G. The sucrose solution side will have more volume.
   H. Both sides will have the same volume.
   J. Not enough information is present to draw a conclusion.

   hint *You can find all the answers for this question in the passage, but if you've done a similar experiment in class, use your knowledge.*

5. Which of the following conclusions can be drawn from the experimental results?

   I.   Osmotic pressure is independent of the solvent used.

   II. Osmotic pressure is only dependent upon the temperature of the system.

   III. Osmosis occurs only when the osmotic pressure is exceeded.

   A. I only
   B. III only
   C. I and II only
   D. I and III only

   hint *First treat each number as an answer choice and check the keywords. Then match your results to the answer choices.*

6. Which of the following figures correctly shows the apparatus used for measuring osmotic pressure, as described in the passage?

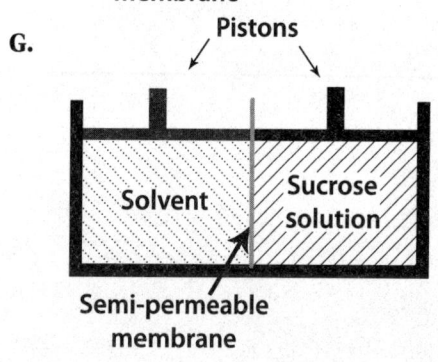

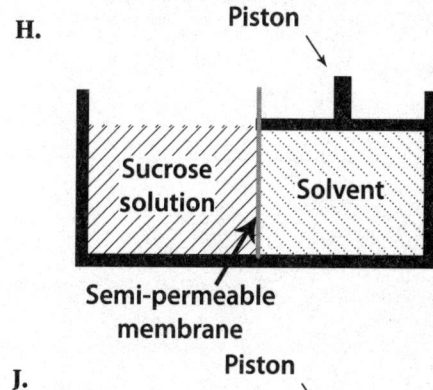

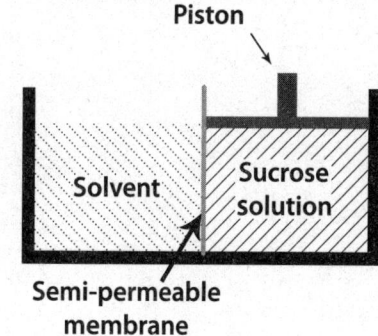

   F. [Diagram: Piston on left side, Sucrose solution (left) | Solvent (right), Semi-permeable membrane]

   G. [Diagram: Pistons on both sides, Solvent (left) | Sucrose solution (right), Semi-permeable membrane]

   H. [Diagram: Piston on right side, Sucrose solution (left) | Solvent (right), Semi-permeable membrane]

   J. [Diagram: Piston on right side, Solvent (left) | Sucrose solution (right), Semi-permeable membrane]

   hint *Begin by matching up diagram labels with keywords from the passage.*

Name _____ Date _____

KAP Wrap

List two ways in which scientific investigation problems are similar to other data passages, and two ways in which they are different.

lesson B
Complex Figure Questions

Thinking KAP

After you have been traveling through the Lacandon Rainforest for several hours, you notice that one of the other tourists is taking notes as she walks. She explains that she is a herpetologist, a scientist who studies reptiles and amphibians. She is collecting information about how many different species of frogs she sees during each hour of the day while you are moving. You volunteer to help her with her research, and she asks you to make a graph of her data at the end of the day.

Time	Number of Species Sighted
8:00–10:00	12
10:00–12:00	8
12:00–2:00	6
2:00–4:00	7
4:00–6:00	13
6:00–8:00	9
8:00–10:00	11

Choose whether to use a line graph, a bar graph, or a scatter plot, then complete the graph below to show how the number of species sighted varied throughout the day.

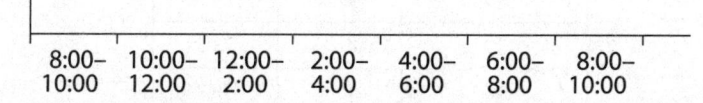

UNIT 3: DATA REPRESENTATION PASSAGES PART 2
LESSON B: COMPLEX FIGURE QUESTIONS

Strategy Instruction

Creating Graphs

Data in the passages will almost always be presented to you in line graphs or tables. However, the ACT also expects you to be familiar with other formats. Because the bar graphs and scatter plots in the questions usually do not have scales marked on their axes, the general shape of the data is very important.

keep in mind

Understanding all of the information will not help you gain points on the test. In fact it will only slow you down. Don't read the text unless you need to. You only earn points on the questions!

Graphing Data

- Determine which values are greater and lesser. Eliminate graphs that do not show these correctly.
- Examine how much greater or how much lesser each data point is in comparison to the ones nearby. Eliminate graphs that do not show these correctly.

Try It Out!

Use the table below to answer the questions on the following page.

The Brazilian tree frog exchanges gases through both its skin and lungs. The exchange rate depends on the temperature of the frog's environment. Fifty frogs were placed in a controlled habitat. The temperature was varied from 5°C to 25°C, and equilibrium was attained before each successive temperature change. The amount of oxygen absorbed by the frogs' lungs and skin per hour was measured, and the results for all the frogs were averaged. The results are shown in Table 1.

Table 1		
Temperature (°C)	Moles O_2 absorbed/hr	
	Skin	Lungs
5	15.4	8.3
10	22.7	35.1
15	43.6	64.9
20	42.1	73.5
25	40.4	78.7

1. Which graph correctly shows the moles of oxygen absorbed per hour by the skin at each temperature?

A. 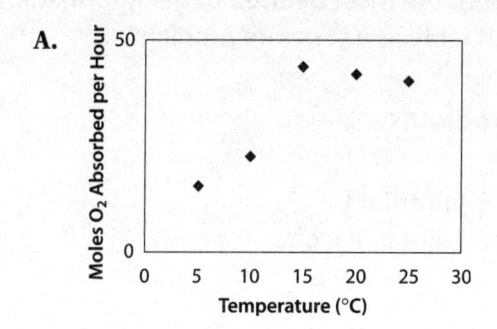
B. 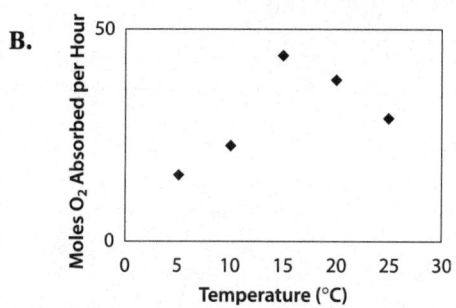
C. 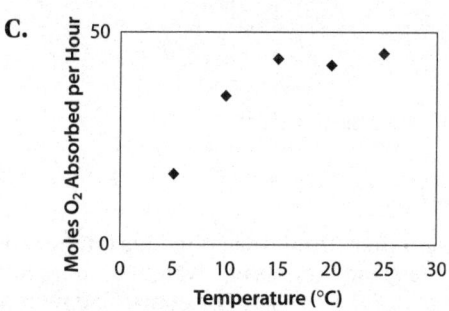
D.

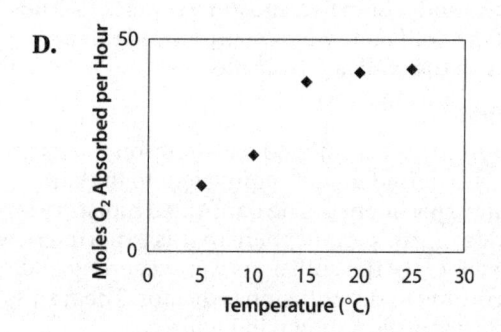

Compare specific data points in each graph to the table and eliminate graphs that do not match.
- Which data point is highest? _____
- Which data point is lowest? _____
- Eliminate each graph that does not match.

How are the remaining graphs different from each other? _____

2. Which of the following graphs shows the moles of oxygen absorbed by the frogs' skin and lungs at each temperature?

F. 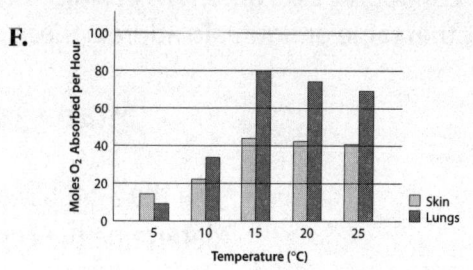
G. 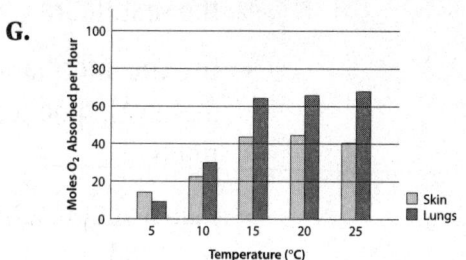
H. 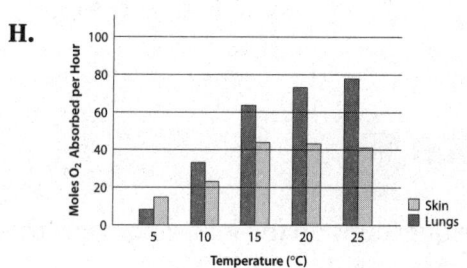
J.

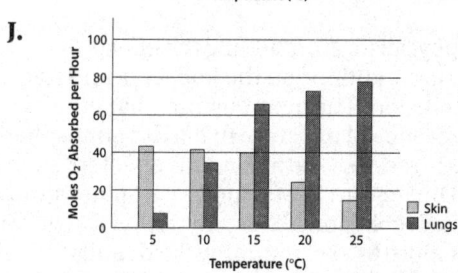

Which graph correctly shows the moles of oxygen absorbed per hour by the skin at each temperature? _____
- Which data point is highest? _____
- Which data point is lowest? _____
- Eliminate each graph that does not match.

How are the remaining graphs different from each other? _____

UNIT 3: DATA REPRESENTATION PASSAGES PART 2
LESSON B: COMPLEX FIGURE QUESTIONS

Using Information from Two Figures

The ACT commonly asks questions in which data from one table or figure must be used to get information from another table or figure. To address these questions, treat each table as a separate problem.

Strategy for Two-Figure Problems

- Use the 3-Step Method for Data Representation to determine the correct data point, trend, or label from the first figure.
- Use the information from the first figure to apply the 3-Step Method for Data Representation to the second figure.
- Match your results to the answer choices.
- Reread the question and answer.

Try It Out!

Read the questions on the following page, then use the figures below to answer them.

Astrophysicists are studying the effects of atmospheric conditions on the impact of an asteroid-to-Earth collision. The most common hypothesis is that the presence of moisture in Earth's atmosphere significantly reduces the hazardous effects of such a collision. One researcher created a computer simulation model of such collisions. This simulation can vary the amount of moisture surrounding Earth and the size of the asteroid. The simulation measures the impact of the collisions using the Torino scale, as shown in Table 1.

Table 1

Torinos	Damage
0 to 0.9	minimal destruction
1 to 3.9	localized destruction
4 to 6.9	regional destruction
7 to 10	global destruction

Experiment 1

The researcher simulated collisions on the Earth model with asteroid models equivalent to mass ranging from 1,000 kg to 3.0×10^{15} kg. The controlled moisture level of the model Earth's atmosphere was 86%. The effects of the collisions were recorded and rated according to the collision indicator.

Experiment 2

The researcher simulated collisions on the Earth model with asteroid models equivalent to the same mass as in Experiment 1. The controlled moisture level of the model Earth's atmosphere in this experiment was 12%. The effects of the collisions were recorded and rated according to the collision indicator. The results of both experiments are shown in Figure 1.

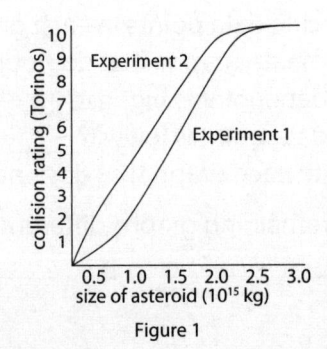

Figure 1

1. If the asteroid that led to the extinction of the dinosaurs 65 million years ago rated 9 Torinos, and Earth's atmospheric moisture level at that time was approximately 86%, what was *most likely* the approximate mass of the asteroid that impacted Earth according to these simulations?
   A. 1.5×10^{15} kg
   B. 2.0×10^{15} kg
   C. 2.5×10^{15} kg
   D. 3.0×10^{15} kg

What do you know about Figure 1?_____

What do you need from Figure 1?_____

What do you know about Figure 2?_____

What do you need from Figure 2? _____

2. In a simulated asteroid-to-Earth collision, a 1.0×10^{15} kg asteroid received a collision rating of 4. The amount of moisture in the atmosphere in the simulation was most likely:
   F. 0%.
   G. 12%.
   H. 86%.
   J. 100%.

What do you know about Figure 1?_____

What do you need from Figure 1?_____

What do you know about Figure 2?_____

What do you need from Figure 2?_____

3. What would be the likely result, according to these simulations, of a 1.5×10^{15} kg asteroid colliding with Earth at an atmospheric moisture level of 12%?
   A. A collision capable of little destruction
   B. A collision capable of localized destruction
   C. A collision capable of regional destruction
   D. A collision capable of global catastrophe

What do you know about Figure 1?_____

What do you need from Figure 1?_____

What do you know about Figure 2?_____

What do you need from Figure 2? _____

Guided Practice

The 3-Step Method for Data Representation

STEP 1: Highlight keywords.

- Read the introduction to determine the general focus of the passage.
- Read a question and highlight the keywords.

STEP 2: Use the diagrams.

- Use the keywords to select the correct headings from the correct diagram.
- Find the data point or summarize the trends.
- If the information is not in the table, read the text above the table.

STEP 3: Answer the question.

- Match the information from the diagram to one of the answer choices.
- Check that the text you have underlined and your analysis of the diagram support your answer.

Use the 3-Step Method for Data Representation to read the passage and answer the questions on the following page.

Researchers studying nutrition disagree on what is the best diet for losing weight. Traditionally, the most recommended diet is one that reduces the number of calories consumed. If a dieter takes in fewer calories than he or she burns, weight loss should eventually result. Dieticians advocating this type of diet often recommend eliminating fat and eating food with a higher proportion of proteins and carbohydrates. This is because fat has nine calories per gram (C/g), whereas proteins and carbohydrates have only four calories per gram.

Recently, some diet experts have started suggesting a different method of weight loss. They recommend placing strict limits on the amount of carbohydrates consumed, while increasing the amount of protein. Dieticians promoting this "low-carb" approach contend that the body readily stores excess carbohydrates, especially when a person is restricting caloric intake, causing the body to act as if it is experiencing a period of famine. Proteins, however, cannot be stored by the human body, and therefore, a diet high in protein will be better for promoting weight loss.

Study 1

Researchers promoting the low-carb approach to dieting set out to show that people lose more weight with this diet than with the traditional low-calorie, low-fat diet. They recruited 40 female nurses working in a major research and teaching hospital. All the nurses in the study were at least slightly overweight, and underwent a medical exam to ensure that they were healthy enough to withstand dieting. Half of the nurses were randomly chosen to follow a low-carb, high-protein diet for six weeks. The remaining half were placed on a reduced-calorie, low-fat diet. The nurses were weighed at the beginning of the study, and once every week thereafter. The results are shown in Table 1.

Table 1		
	Average weight per group (lbs)	
Week:	low-fat	low-carb
0	164.8	165.1
1	163.5	163.7
2	162.2	159.6
3	161.0	157.4
4	159.7	154.2
5	158.4	151.9
6	157.3	149.5

Study 2

Many dieters who experience initial success later gain back some or all of the weight they had lost. The researchers wanted to find out whether the low-carb diet was any different from the conventional diet in terms of long-term weight loss. They again recruited the same nurses who had participated in Study 1. The nurses were not asked to follow any diet plan, but were simply weighed every month for six months after the end of Study 1. The results are shown in Table 2, below.

Table 2		
	Average weight per group (lbs)	
Month	low-fat	low-carb
1	158.8	157.2
2	159.5	163.8
3	160.3	167.5
4	160.7	168.2
5	161.2	169.0
6	161.6	169.4

1. The purpose of Study 1 was to determine whether a low-carb diet or a low-fat diet:
   A. leads to weight gain after six months.
   B. leads to greater weight loss.
   C. causes more calories to be consumed.
   D. has a higher proportion of calories from protein.

2. According to the data in Table 1 and Table 2, on average, the nurses on the low-fat diet, compared to the nurses on the low-carb diet:
   F. lost less weight while they were on the diet, but gained less of it back after six months.
   G. lost less weight while they were on the diet, and gained more of it back after six months.
   H. lost more weight while they were on the diet, and gained less of it back after six months.
   J. lost more weight while they were on the diet, but gained more of it back after six months.

3. Which factor in Study 1 did the researchers manipulate?
   A. The gender of the nurses
   B. The point in time at which the nurses were weighed
   C. The type of diet the nurses were on
   D. The average weight of the two groups of nurses

4. Why do some researchers recommend eating greater amounts of protein and fat, and lower amounts of carbohydrates?
   F. Carbohydrates have more calories per gram than either fat or protein.
   G. Fat has more calories per gram than either protein or carbohydrates.
   H. A low-carbohydrate, high-protein diet results in greater weight loss, even six months after completion of the diet.
   J. Higher protein consumption results in greater weight loss because the calories in protein cannot be stored by the body.

Shared Practice

Use the 3-Step Method for Data Representation to read the passage and answer the questions.

Scientists at an oceanographic station in Alaska have been monitoring ocean conditions to increase understanding of how water conditions are changing over time. These conditions affect local fisheries, a primary economic staple, as well as regional temperature projections.

Experiment 1

Sensors were deployed at nine different depths at the same location on one day. Each sensor recorded the temperature and salinity of the surrounding water. Salinity is the concentration of mineral salts dissolved in water. It may be measured by the mass of total dissolved solids when the water evaporates, electrical conductivity, or osmotic pressure. The measurements in this experiment use electrical conductivity, which is measured in practical salinity units (PSUs) as deviations from a standard. The results are shown in Figure 1.

OCTOBER TEMPERATURE AND SALINITY

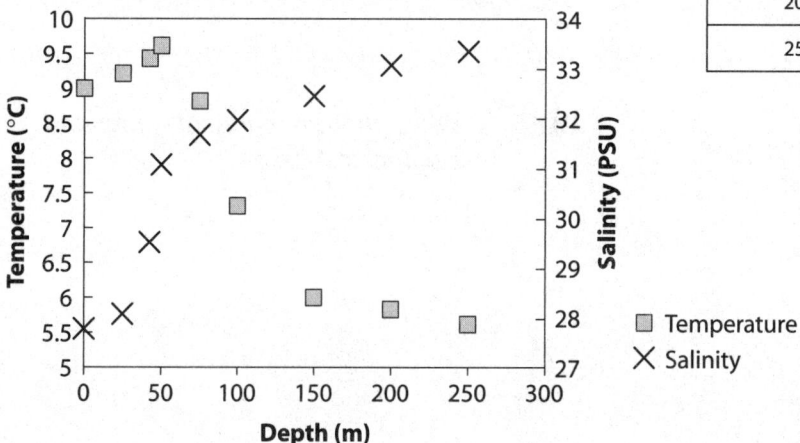

Figure 1

Experiment 2

In an effort to understand some of the reasons behind seasonal fish migration, oceanographers determined the temperature of the ocean at different depths throughout the year. Every day at noon, sensors were lowered to seven different depths and the water temperatures were recorded. The average temperatures for four months are shown in Table 1.

Table 1				
Depth (m)	Average Temperature (°C)			
	January	April	July	October
0	4.8	4.2	13	9
25	4.9	4.1	11	9.2
50	5	4.1	7.5	9.6
100	5.8	4.4	6	7.3
150	6.2	4.8	5.8	6
200	6.4	5	5.8	5.8
250	6.3	5.5	5.8	5.6

1. Before conducting their experiments, the oceanographers hypothesized that temperature and salinity were directly proportional to one another. The data from Figure 1:
   A. support this, because as temperature increases with depth, salinity decreases proportionately.
   B. support this, because as temperature decreases with depth, salinity decreases proportionately.
   C. do not support this, because as salinity increases with depth, temperature decreases proportionately.
   D. do not support this, because although salinity increases with depth, temperature first rises, then falls.

   hint *"Directly proportional" indicates that as one trend increases, the other also increases. "Inversely proportional" indicates that as one trend increases, the other decreases.*

2. In the area of the ocean at which these experiments were done, salinity is directly proportional to the density of the water. As a result, it is likely that:
   F. the density of the water increases with depth.
   G. the density of the water decreases with depth.
   H. the density of the water decreases then increases with depth.
   J. the density of the water increases then decreases with depth.

   hint *Use the keywords to determine which table or figure to collect data from, then summarize the trend.*

3. According to Table 1, as the seasons shift from winter to summer, which of the following is *not* true about the temperature of the ocean?
   A. The surface temperature rises.
   B. The temperature at 250 m drops.
   C. The temperature varies more across the range of depths.
   D. The temperature varies the same amount between January and April as it does from April to July.

   hint *Which months represent winter and summer in Alaska?*

Name _____ Date _____

4. Which of the following is the most accurate graph of how the relationship between temperature and depth changes by season?

F. [graph: Temperature (°C) vs Depth (m), with January, April, July, October data points]

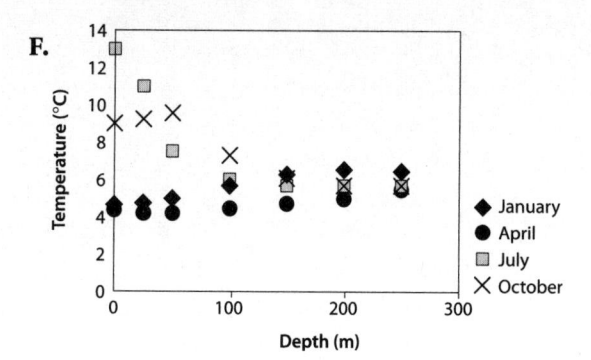

G. [graph: Temperature (°C) vs Depth (m), with January, April, July, October data points]

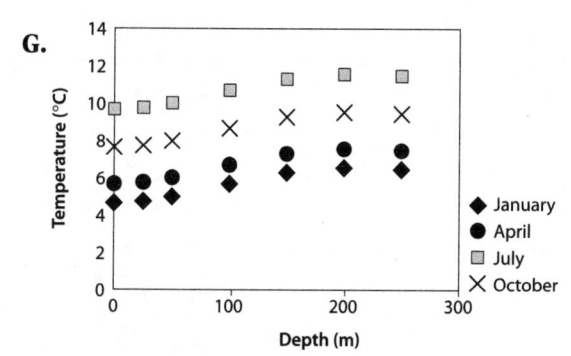

H. [graph: Temperature (°C) vs Depth (m), with January, April, July, October data points]

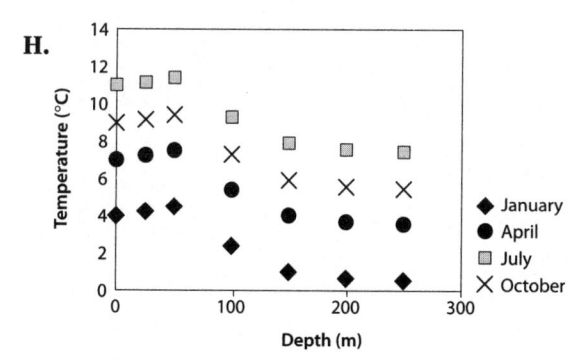

J. [graph: Temperature (°C) vs Depth (m), with January, April, July, October data points]

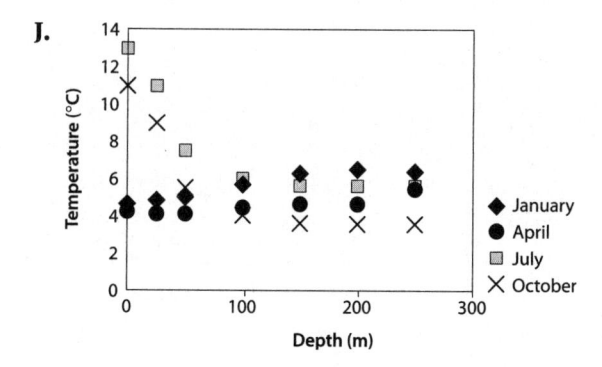

hint *Remember the strategy: find the highest and lowest points or trends first.*

5. The salinity of water at each depth remains constant throughout the year. Assuming that halibut need a temperature of between 8–10°C and a salinity range of 29–30 PSU, during which month will they be found near the research station?

A. January
B. April
C. July
D. October

hint *This question has two parts. First, find the salinity in Figure 1 and use it to find the depth at which halibut swim. Then use Table 1 to determine which month has the correct temperature at that depth.*

KAP Wrap

Which graphing questions are easiest for you? Which ones are hardest? How can you use this information to earn more points on the ACT?

lesson C
Equations in the Science ACT

ReKAP

Review the strategies from Lessons A and B. Then fill in the blanks with what you have learned.

1. When working with tables and figures, first underline the _____, then look for them in the titles, labels, or _____. Then summarize the _____ and look for them in the _____.

2. When working with two experiments, first answer all of the questions about _____, then answer the questions about _____.

3. Some experimental questions require you to predict what will happen in unrelated experiments, explain why something is done, or synthesize a lot of information. You should _____ these problems until the end, then use the passage and your background knowledge to _____ wrong answers, and finally _____ the best one.

Strategy Instruction

Solving Equations

Most ACT exams have one or two questions with equations. These equations use simple addition, subtraction, multiplication, and division, so your main goal is to find the values to substitute in from the tables or text nearby.

Solving an Equation

- Write the equation in a blank space.
- Read the text around the equation and write down what each variable stands for.
- Find the value for each variable and plug it in.
- Determine which answer matches your results.

Try It Out!

Use the following information to answer the questions on the next page.

A high school physics class performed an experiment to determine a snowmobile's total braking distance, or the distance traveled by a snowmobile between when a driver sees a stop sign and when the vehicle comes to a complete halt. After obtaining data, they hypothesized two equations to predict the braking distance, D. To simplify matters, students ignored weather conditions and variations in friction and assumed a constant deceleration.

The class used two methods to predict the total braking distance, D. In Method 1, S is the distance traveled before the driver could begin the braking process when a driver reaction time of 0.8 sec was assumed, and T is the average distance traveled after the brakes were applied. Method 2 assumes that S is negligible and D is twice the initial speed. Table 1 shows the results of both methods with various initial speeds.

Table 1

Initial speed (mi/hr)	Initial speed (ft/s)	Method 1 S (ft)	Method 1 T (ft)	Method 1 D (ft)	Method 2 D (ft)	Experimental Results D (ft)
25	37	30	28	58	74	62
35	51	41	75	116	102	124
45	66	53	144	197	132	202
55	81	65	245	310	162	320

1. Method 2 uses the equation $D = 2 \times I$. If a snowmobile is traveling at an initial speed of 75 ft/s, what will its D be according to Method 2?
   A. 37 ft
   B. 75 ft
   C. 110 ft
   D. 150 ft

Write the equation, then write what each variable stands for.

Substitute in the value of each known variable.

Determine which answer choice matches your results.

2. In Method 1, D can be calculated using the formula:
   F. $S - T$.
   G. $S + T$.
   H. $\dfrac{T}{S}$.
   J. $S \times T$.

Don't try to find the equation on your own. Use the equations they give you and just test out the numbers. The correct one will work with several values. Use the Strategy for Solving an Equation in the space below.

3. The S for an initial speed of 25 mi/hr, in comparison to the S for an initial speed of 55 mi/hr, is approximately:
   A. $\dfrac{1}{4}$ as large.
   B. $\dfrac{1}{2}$ as large.
   C. 2 times as large.
   D. 4 times as large.

Fractions and multiples use another type of equation. Use the Strategy for Solving an Equation in the space below.

Name_____ Date_____

Test Practice Unit 3

When your teacher tells you, carefully tear out this page. Then begin working.

1. Ⓐ Ⓑ Ⓒ Ⓓ   15. Ⓐ Ⓑ Ⓒ Ⓓ

2. Ⓕ Ⓖ Ⓗ Ⓙ   16. Ⓕ Ⓖ Ⓗ Ⓙ

3. Ⓐ Ⓑ Ⓒ Ⓓ   17. Ⓐ Ⓑ Ⓒ Ⓓ

4. Ⓕ Ⓖ Ⓗ Ⓙ

5. Ⓐ Ⓑ Ⓒ Ⓓ

6. Ⓕ Ⓖ Ⓗ Ⓙ

7. Ⓐ Ⓑ Ⓒ Ⓓ

8. Ⓕ Ⓖ Ⓗ Ⓙ

9. Ⓐ Ⓑ Ⓒ Ⓓ

10. Ⓕ Ⓖ Ⓗ Ⓙ

11. Ⓐ Ⓑ Ⓒ Ⓓ

12. Ⓕ Ⓖ Ⓗ Ⓙ

13. Ⓐ Ⓑ Ⓒ Ⓓ

14. Ⓕ Ⓖ Ⓗ Ⓙ

UNIT 3: DATA REPRESENTATION PASSAGES PART 2
LESSON C: EQUATIONS IN THE SCIENCE ACT

SCIENCE TEST
15 Minutes—17 Questions

DIRECTIONS: There are three passages in this test. Each passage is followed by several questions. After reading a passage, choose the best answer to each question and fill in the corresponding oval on your answer sheet. You may refer to the passages as often as necessary.

You are NOT permitted to use a calculator on this test.

Passage I

Indigo buntings (*Passerina cyanea*), night-flying songbirds, are known to migrate each fall to Central America from their breeding grounds in the eastern U.S. They then reverse the journey each spring. Scientists have hypothesized that these birds use the stars to orient themselves during migration. Researchers conducted the following experiments to test this hypothesis.

Experiment 1

Wild, adult *Passerina cyanea* were placed in large, enclosed areas that permitted a full view of the sky. They were observed in April and September, as shown in Figure 1 below.

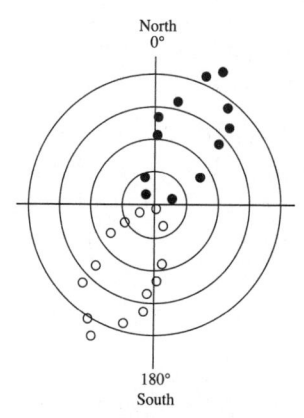

Distance from the origin represents the number of birds observed in the direction indicated, where 0° = True North.

● = April, and ○ = September

Figure 1

Experiment 2

Wild, adult *Passerina cyanea* were placed in a planetarium during April and September. The birds were first exposed to constellations mimicking the night sky, and then to constellations that reversed the normal constellations, putting the northern stars to the south. The results are recorded in Figures 2a and 2b below.

Normal Star Projection

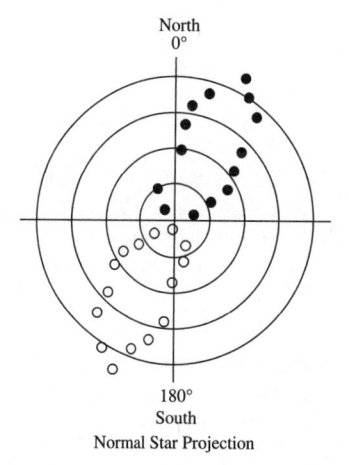

Figure 2a

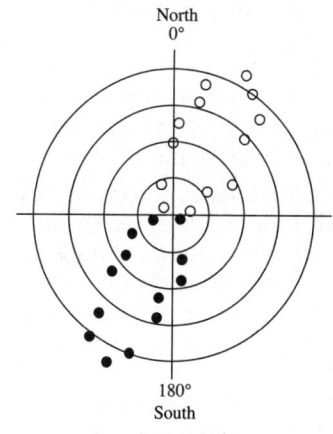

Reversed Star Projection

Figure 2b

GO ON TO THE NEXT PAGE.

Experiment 3

Two groups of newborn indigo buntings (*Passerina cyanea*) were raised by researchers in isolation from adult birds, one group in a planetarium with normal night skies projected, and the other group in a large, windowless area with equal lighting in all areas. Both groups were observed during September in a planetarium with fall constellations projected. The observations are recorded in Figure 3 below.

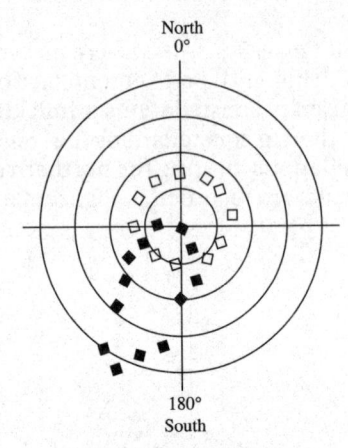

■ = planetarium-raised birds
□ = birds raised in windowless room

Figure 3

1. Which experiments suggest that indigo buntings that are not exposed to the stars cannot orient themselves during migration?
   A. Experiment 1 only
   B. Experiment 2 only
   C. Experiment 3 only
   D. Experiments 2 and 3 only

2. A group of researchers is interested in learning more about how indigo buntings learn to use stars for orientation. Which experiment should they perform next to learn more?
   F. Place adult *Passerina cyanea* in an area with a full view of the sky and ground objects.
   G. Place adult *Passerina cyanea* in a windowless area, then release them.
   H. Raise newborn *Passerina cyanea* in a windowless area with adult *Passerina cyanea*.
   J. Raise newborn *Passerina cyanea* in a windowless area, then allow them to view the stars.

3. If a group of newborn *Passerina cyanea* raised in a windowless area (as in Experiment 3) were observed during April in a planetarium with spring constellations, they would probably display:
   A. no particular orientation.
   B. an orientation similar to that of planetarium-raised birds.
   C. a predominantly eastward orientation.
   D. a predominantly westward orientation.

4. In Experiment 2, the factor that is varied is:
   F. presence of the stars during infancy.
   G. orientation of stars at night.
   H. indoor or outdoor locations.
   J. direction birds flew.

5. Which of the following observations, if true, would weaken the hypothesis that indigo buntings use star positions as a directional guide during migration?

   I. Newborn indigo buntings, raised with adult birds in a windowless area, orient themselves correctly when placed in outdoor cages on clear nights in September.
   II. Indigo buntings in outdoor cages orient themselves correctly during April and September on completely overcast nights.
   III. Other species of migratory birds fly by day when stars are invisible.

   A. I only
   B. II only
   C. II and III only
   D. I and III only

GO ON TO THE NEXT PAGE.

6. Experiment 1 placed wild indigo buntings in an area that allowed them to see the sky. The purpose of this study was to:

- **F.** determine what indigo buntings do under normal conditions.
- **G.** see how indigo buntings are affected by being in a large, enclosed area.
- **H.** observe how indigo buntings teach their children to migrate.
- **J.** protect indigo buntings from predators and hunters in the wild.

GO ON TO THE NEXT PAGE.

Passage II

Milkweed is the common name for the *Asclepias* genus of plants, which vary widely in appearance, and are known for their poisonous, milky white sap. Monarch butterfly larvae have adapted to eat some species of milkweed plants, absorbing the toxins and making the insects foul-tasting or vomit-inducing to predators, who quickly learn not to eat them. Some milkweed plants are so toxic that even Monarch larvae cannot eat them, while some are so mild that they do not protect the Monarchs. Scientists at a national park recently conducted three studies to analyze the relationship between Monarch larvae and the milkweed plants they populate.

Study 1

Scientists compared the traits of two species of milkweed plants, one that is populated by Monarch larvae (Species R), and one that is not populated by Monarch larvae (Species S). The results are shown in Table 1, below.

Table 1		
Milkweed traits	Milkweed species	
	Species R	Species S
Leaves	Thick	Thin
Chemistry	Low pH	High pH
	High concentration of toxins	Low concentration of toxins
Stems	Surrounded by bare ground	Surrounded by weeds and other plants
Blossoms	Purple and fragrant, produce nectar	Not present
Sunlight	Partial sun	Low sun
Larvae per plant	Around 30	Around 3

Study 2

Viceroy butterflies have colors and patterns almost identical to Monarch butterflies, except for a tiny additional black stripe. Their coloring confuses predators, which avoid them because they appear similar to monarchs. Researchers compared the traits of Monarch larvae and Viceroy larvae as shown in Table 2, below.

Table 2		
Larval traits	Larvae	
	Monarch	Viceroy
Coloring	Yellow, white, and black stripes	Green and white patches
Activity	Day	Day and night
Breeding sites	Milkweed leaves	Willow, poplar, and cottonwood leaves

Study 3

Three crops of milkweed were planted outside with several meters separating them. Group 1 consisted of milkweed plants from Species R with no Monarch larvae, Group 2 consisted of milkweed plants from Species R with Monarch larvae placed on them, and Group 3 consisted of milkweed plants from Species S with Monarch larvae placed on them.

All of the plants in the three groups received equal amounts of rainfall and sunlight, and were initially healthy. The number of plants still living in each group was tallied on the first and 60th days of the study. The results are shown in Table 3, below.

Table 3						
Day	Group 1 (Species R plants, no Monarch larvae present)		Group 2 (Species R plants, with Monarch larvae added)		Group 3 (Species S plants, with Monarch larvae added)	
	Alive	Dead	Alive	Dead	Alive	Dead
1	65	0	70	0	68	0
60	30	35	65	5	45	23

GO ON TO THE NEXT PAGE.

7. Using the information from Study 1 and Study 3, approximately how many Monarch larvae, total, were living on the live plants in Group 3 on day 60?
   A. Around 135
   B. Around 69
   C. Around 204
   D. Around 1350

8. In Study 3, which group was used as the control to determine the effects of the variables?
   F. Group 1
   G. Group 2
   H. Group 3
   J. There was no control group.

9. If Viceroy larvae were to eat milkweed plants, they would *most likely*:
   A. prefer Species R plants to Species S.
   B. resemble Monarch butterflies more closely.
   C. be eaten by predators less frequently.
   D. become sick and die.

10. Which graph best represents the results of Study 3?

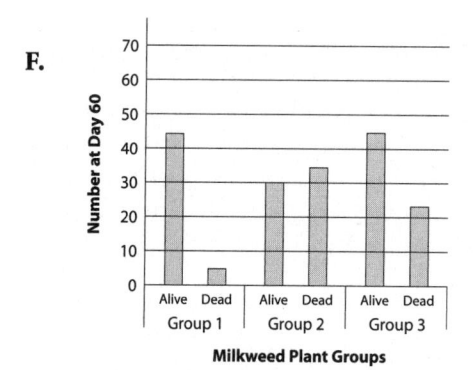

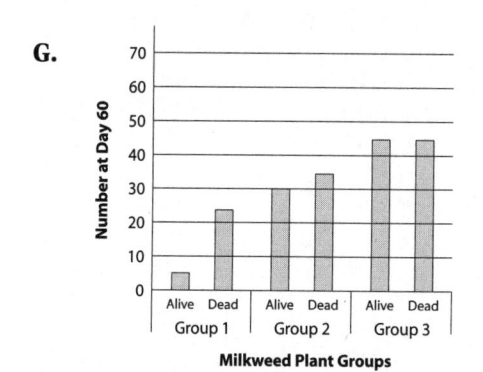

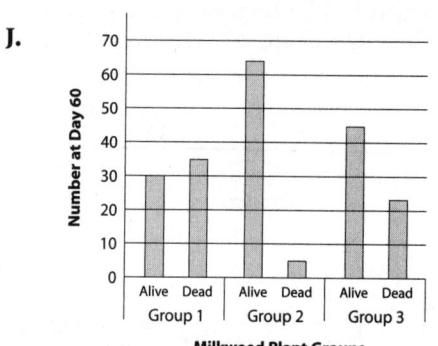

F. [bar graph: Group 1 Alive ~44, Dead ~5; Group 2 Alive ~30, Dead ~34; Group 3 Alive ~44, Dead ~23]

G. [bar graph: Group 1 Alive ~5, Dead ~23; Group 2 Alive ~30, Dead ~34; Group 3 Alive ~44, Dead ~44]

H. [bar graph: Group 1 Alive ~30, Dead ~34; Group 2 Alive ~64, Dead ~14; Group 3 Alive ~29, Dead ~22]

J. [bar graph: Group 1 Alive ~30, Dead ~34; Group 2 Alive ~64, Dead ~5; Group 3 Alive ~44, Dead ~22]

11. The results of Study 3 demonstrate that:
    A. Monarch larvae are necessary for the health of Species R milkweed plants.
    B. Monarch larvae are harmful to the survival of Species R milkweed plants.
    C. Monarch larvae do not affect the growth of Species R milkweed plants.
    D. Monarch larvae use camouflage to disguise Species R milkweed plants from predators.

GO ON TO THE NEXT PAGE.

UNIT 3: DATA REPRESENTATION PASSAGES PART 2
LESSON C: EQUATIONS IN THE SCIENCE ACT

Passage III

A solution is defined as an evenly-distributed (homogeneous) mixture of two or more substances which do not undergo a chemical reaction when mixed. The solvent is the liquid into which the solute, usually a solid or gas, is dissolved. Solutions that can conduct an electrical current are called electrolytes. In such solutions, the current is conducted through an electrolyte by the solute. Solutions that cannot conduct an electric current are called non-electrolytes. The voltage of a current traveling through a solution can be measured by a device called a voltmeter. A simple circuit, consisting of a battery, solution, light-emitting diode (LED), and voltmeter, is diagrammed below in Figure 1.

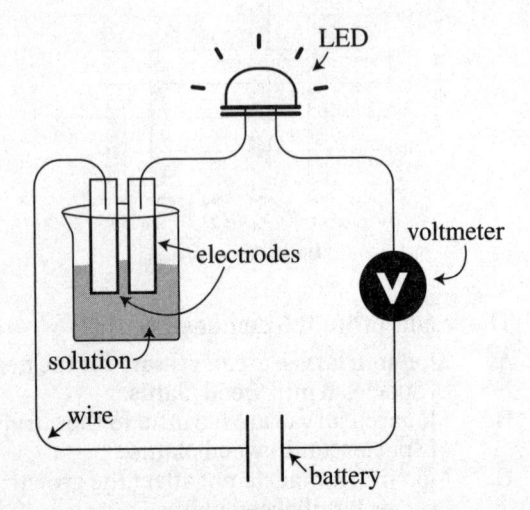

Figure 1

Researchers conducted the following experiments to determine whether increasing the solute in a solution or increasing the temperature of the solution would increase the solution's conductivity.

Experiment 1

Researchers created six solutions, each containing 10.0 grams (g) of a solute dissolved in 100 milliliters (mL) of water at 35°C. Pure water was also tested for conductivity without the addition of a solute. The results of Experiment 1 are shown in Table 1.

Experiment 2

The researchers repeated Experiment 1, this time tripling the amount of solute in each solution to 30.0 g. The solute was again dissolved in 100 mL of water at 35°C. The results of Experiment 2 are shown in Table 1.

Table 1

Solute	Experiment 1: voltmeter reading (V) with 10 g solute	Experiment 2: voltmeter reading (V) with 30 g solute
Water (H_2O)	0.1	0.1
Sodium chloride (NaCl)	5.4	10.6
Sucrose ($C_{12}H_{22}O_{11}$)	0.1	0.1
Potassium fluoride (KF)	5.1	9.7
Magnesium chloride ($MgCl_2$)	1.6	2.6
Acetic acid ($C_2H_4O_2$)	1.9	3.0

Experiment 3

30.0 g of each solute were dissolved in 100 mL of water, and the water was heated to 70°C. The results of Experiment 3 are shown in Table 2.

Table 2	
Solute	Voltmeter reading
Water (H₂O)	0.1
Sodium chloride (NaCl)	1.3
Sucrose (C₁₂H₂₂O₁₁)	0.1
Potassium fluoride (KF)	13.1
Magnesium chloride (MgCl₂)	4.0
Acetic acid (C₂H₄O₂)	4.9

12. Which of the following produces a non-electrolyte when dissolved in water?
    - F. Sodium chloride (NaCl)
    - G. Sucrose ($C_{12}H_{22}O_{11}$)
    - H. Potassium fluoride (KF)
    - J. Acetic acid ($C_2H_4O_2$)

13. If 30.0 g of acetic acid were placed in 100 mL of water and brought to a temperature of 50°C, the conductivity would most likely be:
    - A. 0.1–2.3 V.
    - B. 1.9–3.0 V.
    - C. 3.0V–4.9 V.
    - D. 9.7–13.0 V.

14. Comparing the results of Experiment 2 and Experiment 3 shows that an experimental error was *most likely* made in measuring the voltage of:
    - F. Sodium chloride (NaCl).
    - G. Sucrose ($C_{12}H_{22}O_{11}$).
    - H. Potassium fluoride (KF).
    - J. Magnesium chloride ($MgCl_2$).

15. On the basis of the results in each of the three experiments, which of the following solutions of $MgCl_2$ conducts the most electricity?
    - A. 10g in 100 mL of H_2O at 35°C
    - B. 30g in 100 mL of H_2O at 35°C
    - C. 10g in 100 mL of H_2O at 70°C
    - D. 30g in 100 mL of H_2O at 70°C

16. Assuming no chemical reactions occurred, a solution containing 30 g of sucrose ($C_{12}H_{22}O_{11}$) and 30 g of potassium fluoride (KF) in 100 mL of H_2O at 70°C could be expected to result in a voltmeter reading of approximately:
    - F. 4.9 V.
    - G. 5.1 V.
    - H. 9.7 V.
    - J. 13.1 V.

17. Which factor was varied between Experiment 1 and Experiment 2?
    - A. Volts
    - B. Temperature
    - C. Identity of solutes
    - D. Volume of solutes

END OF TEST.

STOP! DO NOT TURN THE PAGE UNTIL TOLD TO DO SO.

KAP Wrap

Calculators are not allowed on the Science ACT. How will you find the equation that describes your data without one?

Unit 4
Conflicting Viewpoints

lesson A: Questions About One Viewpoint

Thinking KAP

As you are walking through the Lacandon Rainforest, you start getting thirsty. You head toward a stream and are about to take a cool, refreshing drink when one of your friends says, "Don't drink the water! It may be unhealthy."

You stop to ask what could be wrong, when another friend says, "I'm sure it's fine. We're not near any cities, so it can't be polluted."

Which friend do you think is correct? What evidence could you use to demonstrate your opinion?

Strategy Instruction

Introduction to Conflicting Viewpoints

Conflicting Viewpoints passages present two different explanations for the same evidence. Just as two different detectives working on the same case may interpret the evidence in different ways, two scientists may draw different conclusions from the same data. You will be asked to clarify each person's viewpoint, give information that will support or disprove a viewpoint, and compare the ideas of the two people.

> **keep in mind**
> There will only be one Conflicting Viewpoints passage on the test, and it will be followed by seven questions.

The 3-Step Method for Conflicting Viewpoints

STEP 1: Summarize the passage and scan the figures.

- Summarize the introduction carefully and underline keywords.

STEP 2: Answer the easier questions.

- Answer questions about details in the introduction.
- Summarize the first viewpoint and answer questions about the first viewpoint only.
- Summarize the second viewpoint and answer questions about the second viewpoint only.

STEP 3: Answer the harder questions.

- Answer questions about multiple viewpoints.

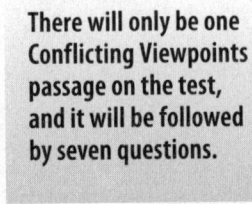

What's in the Introduction?

Sometimes there are questions based solely on the introductory paragraph. Because the viewpoints offer conflicting evidence, the only way you can answer this type of question is to look at the background information in the introduction.

Look for keywords in the introduction which:

- state the conflict to be discussed.
- state evidence that is already known and is agreed upon in both viewpoints.

Try It Out!

Read the introduction below and underline keywords.

Passage I

The formation of geological structures on Earth occurs over long periods of time through the gradual movement of the Earth's crust. One such structure is the Himalayan mountain range in Asia which separates India from the Tibetan plateau. Mineral and fossil evidence shows that the Himalayas began forming approximately 55 million years ago. These studies also show that the mountain range continues to grow from beneath, at the same time that erosion is gradually wearing away its top. Below, two scientists discuss the process behind the creation of mountain ranges such as the Himalayas.

1. According to the passage, the youngest rocks in the Himalayas would most likely be found:
   A. on mountain peaks.
   B. buried deep beneath the mountains.
   C. at the north end of the range.
   D. at the south end of the range.

2. Which of the following statements are both scientists *most likely* to agree with?
   F. The earth formed 55 million years ago.
   G. Mountains are caused by the cooling of Earth's core.
   H. Mountain ranges continue to form today.
   J. The Himalayas separate two tectonic plates.

What problem or question will be discussed in the viewpoints?
_____
_____

One Viewpoint at a Time

Each viewpoint will provide a different explanation for the question or problem presented in the introduction. When you read a viewpoint, underline keywords to answer the following questions:

- What is the writer's theory?
- What evidence does the writer present?

Try It Out!

Underline the theory and evidence for the first viewpoint. Try to underline only 2 or 3 keywords on each line.

Scientist 1

Mountain ranges such as the Himalaya are formed by the process of plate tectonics. The Earth's crust is made up of several massive sheets, known as plates, which float upon a dense and fluid layer known as the mantle. Because the mantle is mobile, the plates above it are constantly in motion. Plates are in contact along their edges, and sometimes the movement of these plates causes one plate to become subducted, or pushed under the other plate. The action of the first plate raises the elevation in the intersecting region, and creates mountain ranges. Because of the gradual movement of these plates, these mountain ranges can continue to increase in elevation or shift noticeably in location.

3. Which of the following models best illustrates the behavior of the Earth's crust on top of the fluid mantle, as described by Scientist 1?
   - **A.** A brick sinks to the bottom of a pool.
   - **B.** Ice cubes float in a glass of water.
   - **C.** Sand moves with a river's current.
   - **D.** Gelatin consolidates from a liquid.

4. Research shows that Mount Everest, a peak within the Himalayan mountain range, is moving northeastward by 2.4 inches (6 cm) per year. Scientist 1 would probably explain this data by saying:
   - **F.** the Indian subcontinental plate is pushing on the Eurasian plate to the west.
   - **G.** the Eurasian plate is pushing on the Indian subcontinental plate to the south.
   - **H.** the cooling of the Earth's core is contracting the Himalayas to the northeast.
   - **J.** the cooling of the Earth's core is contracting the Himalayas to the southwest.

Summarize the writer's theory in one sentence.

_____

_____

The Second Viewpoint

When you read the second viewpoint, treat it as an entirely new idea, and answer all questions that are just about the second viewpoint separately.

Try It Out!

Underline the evidence or reasoning for this writer's theory.

Scientist 2

Mountain ranges like the Himalaya are created through the shrinkage process resulting from a gradual cooling of the Earth. When Earth was forming, it solidified from a state of molten liquid 98 million years ago. During this consolidation process, the temperature of the planet cooled significantly. As the Earth continued to cool, it also reduced in volume. During this process, the mixture of elements also separated into distinct layers, including the outermost layer, known as the crust, which is solid and the least dense. As the inner layers of the planet continued to cool, the subsequent shrinkage in volume resulted in the catastrophic crumpling of the planet's solid surface layer, which created the mountain ranges on the Earth's crust. It is hypothesized that the core of the Earth is still continuing to cool today.

5. According to the information provided, radioactive dating of the Earth's crust would show the rock in the oldest crust described by Scientist 2 to be about how many million years old?
   A. 10
   B. 55
   C. 98
   D. 4,400

6. Scientist 2 would most likely state that the shrinkage in volume of the planet Earth is caused by:
   F. severe changes in the climate.
   G. cooling of the planet's interior.
   H. formation of the Earth's crust.
   J. drifting of the continental plates.

Summarize the writer's theory in one sentence.

_____
_____

Guided Practice

The 3-Step Method for Conflicting Viewpoints

STEP 1: Summarize the passage and scan the figures.

- *Read the introduction carefully and underline keywords.*

STEP 2: Answer the easier questions.

- *Answer questions about details in the introduction.*
- *Summarize the first viewpoint and answer questions about the first viewpoint only.*
- *Summarize the second viewpoint and answer questions about the second viewpoint only.*

STEP 3: Answer the harder questions.

- *Answer questions about multiple viewpoints.*

Use the 3-Step Method for Conflicting Viewpoints to read the passage and answer the questions.

Passage II

The distinction between eukaryotic (nucleated) and prokaryotic (non-nucleated) cells is basic to modern biology. The first prokaryotes appeared two billion years before the first eukaryotes. Most single-celled organisms such as bacteria are prokaryotes, and most complex organisms consist of eukaryotic cells. Eukaryotes contain mitochondria, which are enclosed by inner and outer membranes. It has been suggested that mitochondria-containing eukaryotes evolved from a symbiotic relationship between two types of prokaryotes. Two scientists debate the issue.

Scientist 1

Eukaryotes developed from a symbiotic relationship between two types of prokaryotes. Early prokaryotes did not require O_2; there was no free O_2 in the Earth's atmosphere until certain prokaryotes began releasing O_2 as a metabolic by-product. Eventually, some prokaryotes became aerobic, capable of utilizing free O_2. They were engulfed by anaerobes, prokaryotes that could not metabolize O_2. The aerobes gained a secure environment and a continuous food supply, while the anaerobes gained the ability to survive in an O_2-rich environment. Over time, the symbiotic partners lost their independence, and the aerobic prokaryotes evolved into mitochondria. Mitochondrial DNA differs both genetically and structurally from the DNA in the eukaryotic cell's nucleus.

Scientist 2

Mitochondria developed through invaginations of the cell membrane, creating internal membrane-bound structures similar to vacuoles. The food processing mechanisms of the cell were isolated to this region just as DNA was isolated to the membrane-bound nucleus. Because this process occurred early in the history of the cell before DNA was completely isolated to the nucleus, pertinent DNA remains located in the mitochondria. It should also be noted that although mitochondria synthesize several of the enzymes necessary for their own function, most mitochondrial proteins are controlled by genes in the nucleus of the eukaryotic cell and are synthesized outside of the mitochondria.

1. Scientist 2 mentions that most mitochondrial proteins are controlled by genes in the nucleus in order to:
   A. establish that mitochondria could not have evolved from prokaryotes.
   B. illustrate the superior aerobic capacity of mitochondria.
   C. argue that bacteria are genetically less complex than mitochondria.
   D. demonstrate the prokaryotic nature of mitochondria.

2. Which of the following, if true, would weaken Scientist 1's hypothesis?
   F. Most single-celled organisms are eukaryotes.
   G. Bacteria require many of the same enzymes that mitochondria use in order to reproduce.
   H. The first eukaryotic cells appeared well before the evolution of aerobic prokaryotes.
   J. Random mutations have caused genes to be transferred from the mitochondria to the eukaryotic cell nucleus.

Shared Practice

Use the 3-Step Method for Conflicting Viewpoints to read the passage and answer the questions.

Passage III

Yawning is an involuntary action that involves breathing deeply. At the deepest part of the breath when the lungs have the greatest force, the lungs fill up the rib cage, then lift the ribs, a pattern which coincides with a deep stretch of the jaw and face muscles. The result is a yawn. Yawning has been associated with drowsiness, weariness, or the presence of certain neurotransmitters in the brain. Various medical ailments such as acute myocardial infarction and aortic dissection (vasovagal reactions) may cause excessive yawning. However, the reasons why most humans yawn remain unconfirmed.

Two scientists discuss their views on the causes of yawning.

Scientist 1

Yawning is caused by hypoxia (a depletion of oxygen in the bloodstream). Most yawns occur when a person is tired or bored. At these times, the body is disturbed from a state of homeostasis (an optimal balance of chemicals and functioning in the human body). This disturbance requires an increase in oxygen to return to normal. The respiratory system is stimulated to produce a yawn. The deep breath produced by a yawn suddenly increases the amount of oxygen in the blood and simultaneously rids the body of the excess carbon dioxide that accumulates in the bloodstream because of oxygen deficiency.

Scientist 2

Yawning is caused by a need to stretch in the musculo-skeletal system. Both stretching and yawning most commonly occur during periods of tiredness. Both phenomena increase blood pressure and relieve tension in muscles and joints. Further evidence is found in studies which show that the stretching of facial muscles and the jaw is a necessary part of yawning. Particularly striking support for this theory is found in the behavior of people who are paralyzed on one side of their body from a stroke. It has been observed that such people can stretch limbs on the otherwise paralyzed sides of their bodies when they yawn.

1. There is no correlation between the amount of oxygen in the air and the number of times a person yawns in a day. This statement, if true, would best support the view of:

   A. Scientist 1, because respiration is not a function of yawning.
   B. Scientist 1, because stretching is not a function of yawning.
   C. Scientist 2, because respiration is not a function of yawning.
   D. Scientist 2, because stretching is not a function of yawning.

   hint *Don't answer this question until you've answered all the questions about just Scientist 1 or just Scientist 2.*

2. According to Scientist 2, the best evidence that respiration is not the primary function of yawning is that:
   F. individuals who are paralyzed on one side can stretch limbs on both sides by yawning.
   G. individuals who are paralyzed on one side can stretch their mobile sides by yawning.
   H. most yawns occur when a person is bored or tired.
   J. most people yawn regardless of whether they're tired.

   hint *Highlight the keywords of each answer choice, then look for them in what Scientist 2 has written.*

3. Scientist 1 states that yawning is a body's way of bringing more oxygen into the bloodstream. Which of the following statements, if true, counters this?
   A. A person's blood oxygen levels neither increase nor decrease after a yawn.
   B. A person's blood oxygen levels always increase after a yawn.
   C. A person's blood carbon dioxide levels always decrease after a yawn.
   D. A person's muscles are insufficiently stretched by a yawn.

   hint *This question does not require the passage—just use the information in the question.*

4. Which of the following statements would the two scientists *disagree* about?
   F. Yawns are caused by tiredness or boredom.
   G. Yawns are caused by a deep breath that lifts the rib cage.
   H. Yawns are caused by acute myocardial infarction and aortic dissection.
   J. Yawns are caused by a need to stretch.

   hint *Most statements the scientists agree about will be in the introduction. Underline keywords in each answer choice, then look for them in the introduction.*

Name_____ Date_____

5. According to Scientist 1, which of the following would keep a person from yawning?
   A. Increase the amount of carbon dioxide in the bloodstream.
   B. Stretch throughout the day.
   C. Reduce pressure on the lungs.
   D. Increase the amount of oxygen in the bloodstream.

   hint ▷ This question only asks about Scientist 1. Answer this question before reading about Scientist 2.

6. According to the passage, excessive or abnormal yawning is associated with:
   F. watching other people yawn.
   G. insufficient lung capacity.
   H. drowsiness or weariness.
   J. acute myocardial infarction.

   hint ▷ This answer is in the introduction. Answer this question first.

UNIT 4: CONFLICTING VIEWPOINTS
LESSON A: QUESTIONS ABOUT ONE VIEWPOINT

KAP Wrap

The 3-Step Method for Conflicting Viewpoints tells you to answer questions about just one viewpoint before answering questions about both. Why is this a better idea than answering the questions in the order they are printed?

Lesson B: Analyzing the Passage

Thinking KAP

As you are walking through the Lacandon Rainforest later in the day, you tell your friends about the herpetologist in the group and explain that she studies reptiles and amphibians. You explain that you made a graph for her that showed that there were many more species of frogs sighted in the morning and evening than around noon.

One friend suggests that the frogs do not like the hot sun, and hide in the shade during the afternoon. Another suggests that your tour group was making more noise in the afternoon and scared the frogs away. A third friend suggests that as you walked through the rainforest you entered areas with different amounts of water, sunlight, and insects, and some of them held more frogs than others.

Which friend do you think is correct? What evidence could you use to demonstrate your opinion?

Strategy Instruction

Evaluating Evidence

As you learned in Lesson A, each viewpoint contains a theory and supporting evidence. Although easier questions will ask you to identify details from the viewpoints, harder questions may ask you to speculate on additional evidence that would weaken or strengthen a theory.

There are three main types of harder questions on the ACT. These questions will ask you to:

- determine whether new evidence supports or contradicts a theory.
- predict the results of an experiment.

In all of these cases, Evaluating Evidence can be used to find exactly the information you need and contrast it with the evidence in the question.

keep in mind

Keywords may appear in the questions or the answers.

Evaluating Evidence

- Read the experiment or results and underline keywords.
- Look for similar keywords in the viewpoint, and reread the sentence.
- Determine whether the ideas in the question are the same as or the opposite of the ideas in the viewpoint.

Supporting and Contradicting Data

Scientific theories are based upon support from many experiments. These same experiments can be used to disprove other competing theories. Many ACT questions will ask you to determine whether a specific piece of evidence supports or weakens the ideas presented in the viewpoints. To do this, use Evaluating Evidence to address each piece of new evidence.

Try It Out!

Read the viewpoint below. Underline the theory and the evidence that support it. Remember to underline only two or three words per line.

Passage I

The laws governing the movement of objects through space have been debated by many scientists throughout history. Two scientists discuss how objects in motion behave over time.

Scientist 1

Objects which are in motion remain in motion at the same speed, unless acted upon by an outside force. The moon and planets continue to remain in motion at a constant speed because, in the vacuum of space, nothing acts upon them to slow them down or cause them to change direction. However, if an object is rolled across a surface, the surface exerts a force that acts against the object to slow it down and eventually causes it to stop. The rougher the surface, the more force exerted.

1. Which of the following findings, if true, could be used to *weaken* Scientist 1's argument?
    A. Objects on Earth slow down at predictable rates.
    B. Objects speed up when rolled down a hill.
    C. Rougher surfaces allow objects to travel farther.
    D. Rougher surfaces cause objects to stop sooner.

Is (A) in the first viewpoint? _____ If so, is it the same or opposite? _____

Is (B) in the first viewpoint? _____ If so, is it the same or opposite? _____

Is (C) in the first viewpoint? _____ If so, is it the same or opposite? _____

Is (D) in the first viewpoint? _____ If so, is it the same or opposite? _____

Compare and Contrast

All of the questions that apply to one scientist can also be applied to two scientists. These questions will either ask you to determine which scientist supports a specific idea, or identify which theory would be strengthened or weakened by a new piece of evidence.

Try It Out!

Scientist 2

   Objects moving on Earth are governed by different laws than planetary motion. The moon and planets continue to move at the same speed and in predictable directions because they are governed by unique laws and are unlike objects on Earth. Objects on Earth, on the other hand, can be observed to decelerate and eventually stop. No case has ever been demonstrated in which an object on Earth continued moving indefinitely.

2. A scientist gives evidence that the Earth's spin is slowing down by about 1.5–2 milliseconds per century. If true, this statement *weakens* the hypothesis of:

   F. Scientist 1 only.
   G. Scientist 2 only.
   H. both Scientist 1 and Scientist 2.
   J. neither Scientist 1 nor Scientist 2.

What does this new evidence tell you about the motion of objects in space? _____

_____

Does this evidence support or weaken the first theory? _____

Does this evidence support or weaken the second theory? _____

Predicting Results

Scientific theories are useful to scientists because they not only explain the causes of experiments or data that are already available, but help predict the outcome of new experiments. Evaluating Evidence can be used for predicting the results of experiments as well. Simply treat each prediction as if it were evidence, and compare it to the evidence in the viewpoint.

Try It Out!

Use the viewpoints on the previous two pages to answer the questions below using Evaluating Evidence. Remember to underline keywords to help you look.

3. A scientist observes that when a marble is rolled down one smooth slope and up another, it reaches the same total height, even if the second slope is at a different angle, as shown in Figures 1 and 2 below. If nothing gets in the marble's way, what would each scientist predict would happen to the marble in Figure 3?

   A. Scientist 1 would predict that the marble would continue moving forever.
   B. Scientist 2 would predict that the marble would continue moving forever.
   C. Both Scientists would predict that the marble would continue moving forever.
   D. Neither Scientist would predict that the marble would continue moving forever.

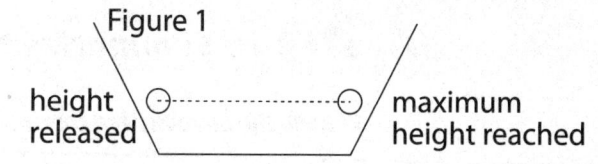

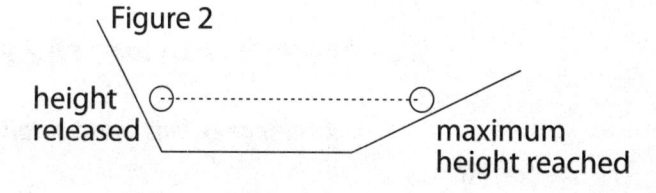

What does Scientist 1 predict about objects traveling along surfaces? _____

_____

What does Scientist 2 predict about objects traveling along surfaces? _____

_____

UNIT 4: CONFLICTING VIEWPOINTS
LESSON B: ANALYZING THE PASSAGE

159

Guided Practice

The 3-Step Method for Conflicting Viewpoints

STEP 1: Summarize the passage and scan the figures.

- *Read the introduction carefully and underline keywords.*

STEP 2: Answer the easier questions.

- *Answer questions about details in the introduction.*
- *Summarize the first viewpoint and answer questions about the first viewpoint only.*
- *Summarize the second viewpoint and answer questions about the second viewpoint only.*

STEP 3: Answer the harder questions.

- *Answer questions about multiple viewpoints.*

Use the 3-Step Method for Conflicting Viewpoints to read the passage and answer the questions.

Passage II

The evolutionary origin of the class *Aves*, or modern-day birds, is still under debate. Two scientists discuss whether birds are descendants of dinosaurs or other reptiles.

Scientist 1

Modern-day birds evolved from dinosaur ancestors. Skeletal structures of dinosaurs and birds share many similarities. Recent fossil findings show that certain dinosaurs and birds both have fused bones in the chest, known as the furcula, or wishbone, which is not present in other animals. These dinosaurs, known as theropods, and birds also share a common foot structure, and the presence of a small bone at the ankle known as the astralagus. Furthermore, the discovery of several fossils, such as *Archaeopteryx* in Germany and the *Sinosauropteryx* in China show an outer covering that resembles feathers or stiff, hair-like "proto-feathers." These skeletons are difficult to classify definitively as dinosaurs or birds, indicating an evolutionary relationship between these two groups.

Scientist 2

Birds evolved from an early family of reptiles which are likely to include the fossil findings of *Protoavis* and *Archaeopteryx*. These reptiles would have been arboreal (living in trees), because most ancestors of flying species are arboreal. Skeletal studies of primitive bird fossils also reveal that the structure of the rib cage and the respiratory apparatus of birds is significantly different from modern reptiles, which may indicate a much earlier divergence of evolutionary paths, before the presence of dinosaurs. In fact, radiocarbon dating of fossils reveals problems for current theories regarding the evolutionary pathway of birds. Many proposed ancestors occur too late in the records to be reasonably related to modern-day birds.

1. The presence of the semi-lunate carpal bone is noted in the wrist joints of birds, but not in those of the majority of theropods. Based on the information provided, this finding would most likely weaken the viewpoint(s) of:
   A. Scientist 1 only.
   B. Scientist 2 only.
   C. both Scientist 1 and Scientist 2.
   D. neither Scientist 1 nor Scientist 2.

2. Which of the following would *weaken* the argument of Scientist 1?
   F. The discovery of animals with proto-feathers from before the time of dinosaurs
   G. The discovery of reptiles from before the time of dinosaurs
   H. The discovery of animals living in trees from before the time of dinosaurs
   J. The discovery of fossils from before the time of dinosaurs

Shared Practice

Use the 3-Step Method for Conflicting Viewpoints to read the passage and answer the questions.

Passage III

Will the universe continue to expand or will it eventually collapse? The answer to this question depends on the average density of all matter in the universe. If it is greater than the critical density of 6×10^{-27} kg/m^3, then the universe is "closed" and it will eventually stop expanding and begin contracting. If the average density is less than this critical value, then the universe is "open" and the expansion will continue.

Two scientists discuss calculation of the average density of matter in the universe.

Scientist 1

Astronomers can estimate the average density of the universe by tabulating all the detectable matter over a large volume of the universe. The mass of galaxies, intergalactic stars, and gas has been determined from luminosity (brightness) measurements. From these measurements, the average density of the universe was found to be only 3×10^{-28} kg/m^3 or 5%, (or at most 10%) of the critical value. Therefore, the universe is "open" and will continue to expand.

Scientist 2

There is a great deal more mass in the universe than has been detected. Observations of the motions of stars in other galaxies indicate that the force of gravity is greater than that which the total mass of detected matter could possibly exert. The missing mass exerting this gravitational pull is known as non-luminous, or dark, matter which cannot be detected by luminosity measurements. Both theories and observations based on the gravitational force exerted indicates that over 90% of matter in the universe (as much as 99%) is dark matter. When calculations of average density are estimated taking this other matter into consideration, the average density of the universe does exceed the critical value. Therefore, the universe must be "closed"—it will stop expanding, begin contracting, and eventually collapse.

1. Scientist 1 and Scientist 2 disagree on the point that:
   A. the universe will be closed if the mass is greater than the critical density.
   B. luminosity measurements are related to brightness.
   C. the majority of the matter in the universe is currently undetectable.
   D. the amount of luminous matter alone is insufficient for an eventual contraction.

hint *Remember, things that scientists agree about are almost always in the introduction.*

Name_____   Date_____

2. Both scientists agree that:
   F. the universe is currently expanding.
   G. the universe is currently static.
   H. the universe is currently contracting.
   J. critical mass cannot be determined.

   hint *The keyword in this question is "currently." Begin by reading the introduction to see what both scientists agree is "currently" happening.*

3. Which further experiment would help to support the hypothesis of Scientist 2?
   A. Measuring the mass of all non-luminous matter in the universe
   B. Showing that the universe's average density is exactly 27% of the critical value
   C. Determining the origin of the universe
   D. Developing a method for measuring the temperature at which galaxies come together

   hint *Eliminate answers that don't make sense.*

4. According to Scientist 1, the average density of the universe can be determined by:

   I. measuring luminosities.
   II. studying the motions of planets in galaxies.
   III. determining the mass of non-luminous matter.

   F. I only
   G. II only
   H. I and III only
   J. II and III only

   hint *For number questions, check each number individually, and then look at the answer choices.*

Name_____ Date_____

5. The evidence presented by Scientist 1 supports which of the following conclusions?
   A. The universe is much closer to the critical density than current calculations indicate.
   B. The critical density is 10 to 20 times greater than the calculated average density of the universe.
   C. The total mass of stars cannot be determined from their brightness.
   D. The "non-luminous matter" of stars accounts for about 5% of the total mass of the universe.

   hint *Use keywords to find the statement that matches the passage.*

6. Recent evidence suggest that tiny particles called neutrinos may have mass, but are not detectable through measurements of luminosity. If this is true, it best supports:
   F. Scientist 1 only.
   G. Scientist 2 only.
   H. both Scientist 1 and Scientist 2.
   J. neither Scientist 1 nor Scientist 2.

   hint *Use keywords to match the evidence with the appropriate theory.*

KAP Wrap

What makes a Conflicting Viewpoints question hard? What strategies do you know for answering hard Conflicting Viewpoints questions?

lesson C

General Test-Taking Strategies

ReKAP

Review the strategies from Lessons A and B, then fill in the blanks with what you have learned.

1. Put the following steps in order from 1–7 for addressing passages with two viewpoints.

   ___ answer questions comparing both viewpoints

   ___ read the introduction

   ___ read the second viewpoint

   ___ answer questions about the second viewpoint only

   ___ read the first viewpoint

   ___ answer questions about the introduction

   ___ answer questions about the first viewpoint only

2. When solving problems that involve evidence, first _____

   in the question, then compare them to specific _____ in the viewpoint.

   Finally, determine if the pieces of evidence are _____ or
   _____.

Strategy Instruction

Relax!

When you take the ACT Science test, you will need to work for 40 minutes the first day, and for an hour the second day. Just as a marathon runner drinks water every few miles to keep going, you will need a short break every now and then.

Every few pages, you should take a silent stretch. The Silent Stretch takes about 15 seconds and makes you feel refreshed.

keep in mind

Take a stretch after every few pages. You don't need to stretch after every page.

The Silent Stretch

- While you count slowly to 15, stretch your shoulders back.
- At the same time, stretch your arms down toward the floor as far as they will go.
- At the same time, stretch your legs straight out in front of you, keeping them close to the floor.
- Take deep breaths as you count. Breathe in, 1...2...3, breathe out, 4...5...6, breathe in, 7...8...9, breathe out, 10...11...12—all the way to 15.

You should learn to do this so quietly that no one in the room notices you doing it. When you open your eyes and go back to the test after the Silent Stretch, you will feel refreshed!

Get Ready

What you don't want to do right before a big test is cram. You've learned a lot of science this year. You've also learned a lot of test-taking strategies. You are totally prepared, and you don't need to cram. You also don't need to worry when you are taking the test—just remember the 3-Step Method for ACT Science!

Try It Out!

Read each statement below. Circle DO or DON'T for each statement.

1. The night before the test, make sure you review all the science facts you can find.   DO   DON'T

2. Eat a healthy dinner the night before the test.   DO   DON'T

3. On the morning of the test, skip breakfast to get in some extra studying.   DO   DON'T

4. Right before you take the test, make a list of all the things you need to do that day.   DO   DON'T

5. Answer all the test questions in order.   DO   DON'T

6. Answer the easiest questions first.   DO   DON'T

keep in mind

If you are worried about the test, talk to your family or your teacher about it.

Test Practice Unit 4

When your teacher tells you, carefully tear out this page. Then begin working.

1. Ⓐ Ⓑ Ⓒ Ⓓ

2. Ⓕ Ⓖ Ⓗ Ⓙ

3. Ⓐ Ⓑ Ⓒ Ⓓ

4. Ⓕ Ⓖ Ⓗ Ⓙ

5. Ⓐ Ⓑ Ⓒ Ⓓ

6. Ⓕ Ⓖ Ⓗ Ⓙ

7. Ⓐ Ⓑ Ⓒ Ⓓ

8. Ⓕ Ⓖ Ⓗ Ⓙ

9. Ⓐ Ⓑ Ⓒ Ⓓ

10. Ⓕ Ⓖ Ⓗ Ⓙ

11. Ⓐ Ⓑ Ⓒ Ⓓ

12. Ⓕ Ⓖ Ⓗ Ⓙ

13. Ⓐ Ⓑ Ⓒ Ⓓ

14. Ⓕ Ⓖ Ⓗ Ⓙ

UNIT 4: CONFLICTING VIEWPOINTS
LESSON C: GENERAL TEST-TAKING STRATEGIES

SCIENCE TEST
15 Minutes—14 Questions

DIRECTIONS: There are three passages in this test. Each passage is followed by several questions. After reading a passage, choose the best answer to each question and fill in the corresponding oval on your answer sheet. You may refer to the passages as often as necessary.

You are NOT permitted to use a calculator on this test.

Passage I

Peptic ulcers, or ulcers of the stomach, are suspected if a patient complains of recurrent abdominal pain, heartburn, or gastroesophageal reflux disease (GERD). Ulcers are painful because they are sores, or patches of the digestive tract lining which are damaged and become more sensitive to the action of highly acidic secretions of the stomach, such as gastric acid.

Ulcers are diagnosed through a test known as esophagogastroduodenoscopy, or EGD, in which the interior of the digestive tract is examined for these sensitive regions using a tool known as an endoscope. An endoscope is a camera and fiberoptic light system mounted on a thin tube which can be inserted down the digestive tract. Ulcers can be diagnosed using the pictures taken using an endoscope.

Two scientists discuss causes of ulcers.

Scientist 1

Peptic ulcers are caused by physical and/or mental stress. An individual who is experiencing painful symptoms associated with an ulcer can take antacids to neutralize stomach acid's low pH. However, this will only temporarily relieve the discomfort of the ulcer without healing it. The only treatment that will cure the ulcer is to modify one's lifestyle in order to minimize the causes of stress. Some of these causes include smoking, spicy foods, and emotional stress. Some individuals of certain blood types may also be more prone to develop ulcers as well.

Scientist 2

Peptic ulcers are caused by the presence of a bacteria known as *Helicobacter pylori* (*H. pylori*), which can live in acidic environments such as those found in parts of the digestive tract. When *H. pylori* is present in the digestive tract, it indirectly stimulates an increased production of gastric acid. Treatment of an ulcer involves administering antibiotics such as ampicillin, which will destroy the *H. pylori* population in the digestive tract. Once the bacterial infection is gone, the symptoms resolve and the lining of the digestive tract is able to heal.

GO ON TO THE NEXT PAGE.

1. A 12-year study done with women shows that smokers are almost twice as likely to have an ulcer than non-smokers. Based on the information provided, this finding would most likely strengthen the viewpoint(s) of:
   A. Scientist 1 only.
   B. Scientist 2 only.
   C. both Scientist 1 and Scientist 2.
   D. neither Scientist 1 nor Scientist 2.

2. Both scientists would most likely agree that the painful symptoms of an ulcer are caused by:
   F. prolonged mental, physical, and emotional stress.
   G. an infection of *H. pylori* in the digestive tract.
   H. a diagnostic procedure using an endoscope.
   J. the sensitivity of damaged stomach lining to gastric acid.

3. *Helicobacter pylori* is the only known organism that is capable of surviving acidic environments with pH measurements as low as that of the stomach. Species that are closely related to *H. pylori* have been found in the human liver as well. Based on the information provided, this finding would most likely weaken the viewpoint(s) of:
   A. Scientist 1 only.
   B. Scientist 2 only.
   C. both Scientist 1 and Scientist 2.
   D. neither Scientist 1 nor Scientist 2.

4. The information in the passage indicates that a diagnosis of an ulcer is made using:
   F. antibiotics.
   G. stomach acid.
   H. endoscopy.
   J. antacids.

5. Which scientist(s), if either, would suspect an ulcer in a person whose symptoms included heartburn?
   A. Scientist 1 only
   B. Scientist 2 only
   C. Both Scientist 1 and Scientist 2
   D. Neither Scientist 1 nor Scientist 2

6. According to Scientist 1, which of the following situations is most likely to cause an ulcer?
   F. An infection of *H. pylori* in the stomach
   G. Regular consumption of foods high in fat
   H. Light to moderate levels of exercise
   J. A 20-year history of smoking

7. According to Scientist 2, which of the following stages is not part of the healing process?
   A. Mending of the digestive tract lining
   B. Making adjustments to lifestyle and diet
   C. Destruction of *H. pylori* in the stomach
   D. Relief of painful abdominal discomfort

GO ON TO THE NEXT PAGE.

Passage II

While the focus (point of origin) of most earthquakes lies less than 20 km below Earth's surface, certain seismographic readings indicate that some activity originates at considerably greater depths. Below, two scientists explain the possible causes of deep-focus earthquakes.

Scientist 1

Deep-focus earthquakes are caused by the pressure of fluids trapped in Earth's tectonic plates. Below 50 km, rock is under too much pressure to fracture normally. Instead, as a plate is forced down into the mantle by convection, increases in temperature and pressure cause changes in the crystalline structure of minerals found in the rock. In adopting a denser configuration, the crystals dehydrate, releasing water. Other sources of fluid include water trapped within the pockets formed by deep-sea trenches, which are carried down with the shifting of plates. Laboratory work has shown that fluids trapped in rock pores can cause rock to fail at lower shear stresses. In fact, at the Rocky Mountain Arsenal, the injection of fluid wastes into the ground accidentally induced a series of shallow-focus earthquakes.

Scientist 2

Deep-focus earthquakes result from the slippage that occurs when rock in a descending tectonic plate undergoes a phase change in its crystalline structure along a plane parallel to a stress. These quakes cannot result from normal fractures because rock becomes ductile at temperatures and pressures that exist at depths greater than 50 km. Such phase changes have been produced in the laboratory by compressing a slab of calcium magnesium silicate. Furthermore, in most seismic zones, the recorded incidence of deep-focus earthquakes corresponds to the depths at which phase changes are predicted to occur in mantle rock. The majority of phase changes will occur between 400 km and 700 km below the surface, and deep-focus earthquake seismic activity is negligible above 400 km, significant between 400 and 680 km, and not recorded below 700 km.

8. Scientists 1 and 2 agree on which point?
   F. Deep-earthquake activity does not occur below 400 km.
   G. Fluid allows tectonic plates to slip past one another.
   H. Water can penetrate mantle rock.
   J. Rock below 50 km will not fracture normally.

9. Which of the following is evidence that would support the hypothesis of Scientist 1?
   A. The discovery that water can be extracted from mantle-like rock at temperatures and pressures similar to those found below 300 km
   B. Seismographic indications that earthquakes occur 300 km below the surface of Earth
   C. The discovery that phase changes occur in the mantle rock at depths of 1,080 km
   D. An earthquake underneath Los Angeles that was shown to have been caused by water trapped in sewer lines

10. Both scientists assume that:
    F. deep-focus earthquakes are more common than surface earthquakes.
    G. phase changes in the crystalline structure of rocks cause surface earthquakes.
    H. Earth's crust is composed of mobile tectonic plates.
    J. deep-focus earthquakes cannot be felt on Earth's crust without special recording devices.

GO ON TO THE NEXT PAGE.

11. Which of the following would weaken Scientist 2's hypothesis?
    A. Finding evidence of other sources of underground water
    B. Recording a deep-focus earthquake below 680 km
    C. Finding a substance that doesn't undergo phase changes even at depths equivalent to 680 km
    D. Demonstrating that rock becomes ductile at depths of less than 50 km

12. According to Scientist 1, the earthquake at Rocky Mountain Arsenal occurred because:
    F. serpentine or other minerals dehydrated and released water.
    G. fluid wastes injected into the ground compressed a thin slab of calcium magnesium silicate.
    H. fluid wastes injected into the ground flooded pockets of deep-sea trench.
    J. fluid wastes injected into the ground lowered the shear stress failure point of the rock.

13. Scientist 2's hypothesis would be strengthened by evidence showing that:
    A. water evaporates at high temperatures and pressures.
    B. deep-focus earthquakes can occur at 680 km.
    C. stress has the same effect on mantle rock that it has on calcium magnesium silicate.
    D. water pockets exist at depths below 300 km.

14. The information in the passage indicates that:
    F. most earthquakes originate near Earth's surface.
    G. most earthquakes originate hundreds of kilometers below Earth's surface.
    H. deep-sea earthquakes often are caused by volcanic activity.
    J. earthquakes are Earth's most destructive natural phenomenon.

END OF TEST.

STOP! DO NOT TURN THE PAGE UNTIL TOLD TO DO SO.

Name _____ Date _____

KAP Wrap

The ACT gives you 35 minutes to answer 40 questions. What strategies and methods can you use to help organize your time so you get the most points?

Practice Test 2

Practice Test 2

Today, you will take the second Practice Test. Like the one you took at the beginning of Unit 1, this test will replicate the format and test conditions of the science ACT.

You will see seven passages on this Practice Test. Each passage will be followed by 5–7 questions. The ACT does not deduct points for incorrect answers, so if you are unsure about a question, just take your best guess. Remember to use all the methods and strategies you've learned!

Practice Test 2

When your teacher tells you, carefully tear out this page. Then begin working.

1. Ⓐ Ⓑ Ⓒ Ⓓ
2. Ⓕ Ⓖ Ⓗ Ⓙ
3. Ⓐ Ⓑ Ⓒ Ⓓ
4. Ⓕ Ⓖ Ⓗ Ⓙ
5. Ⓐ Ⓑ Ⓒ Ⓓ
6. Ⓕ Ⓖ Ⓗ Ⓙ
7. Ⓐ Ⓑ Ⓒ Ⓓ
8. Ⓕ Ⓖ Ⓗ Ⓙ
9. Ⓐ Ⓑ Ⓒ Ⓓ
10. Ⓕ Ⓖ Ⓗ Ⓙ
11. Ⓐ Ⓑ Ⓒ Ⓓ
12. Ⓕ Ⓖ Ⓗ Ⓙ
13. Ⓐ Ⓑ Ⓒ Ⓓ
14. Ⓕ Ⓖ Ⓗ Ⓙ

15. Ⓐ Ⓑ Ⓒ Ⓓ
16. Ⓕ Ⓖ Ⓗ Ⓙ
17. Ⓐ Ⓑ Ⓒ Ⓓ
18. Ⓕ Ⓖ Ⓗ Ⓙ
19. Ⓐ Ⓑ Ⓒ Ⓓ
20. Ⓕ Ⓖ Ⓗ Ⓙ
21. Ⓐ Ⓑ Ⓒ Ⓓ
22. Ⓕ Ⓖ Ⓗ Ⓙ
23. Ⓐ Ⓑ Ⓒ Ⓓ
24. Ⓕ Ⓖ Ⓗ Ⓙ
25. Ⓐ Ⓑ Ⓒ Ⓓ
26. Ⓕ Ⓖ Ⓗ Ⓙ
27. Ⓐ Ⓑ Ⓒ Ⓓ
28. Ⓕ Ⓖ Ⓗ Ⓙ

29. Ⓐ Ⓑ Ⓒ Ⓓ
30. Ⓕ Ⓖ Ⓗ Ⓙ
31. Ⓐ Ⓑ Ⓒ Ⓓ
32. Ⓕ Ⓖ Ⓗ Ⓙ
33. Ⓐ Ⓑ Ⓒ Ⓓ
34. Ⓕ Ⓖ Ⓗ Ⓙ
35. Ⓐ Ⓑ Ⓒ Ⓓ
36. Ⓕ Ⓖ Ⓗ Ⓙ
37. Ⓐ Ⓑ Ⓒ Ⓓ
38. Ⓕ Ⓖ Ⓗ Ⓙ
39. Ⓐ Ⓑ Ⓒ Ⓓ
40. Ⓕ Ⓖ Ⓗ Ⓙ

SCIENCE TEST

35 Minutes—40 Questions

DIRECTIONS: There are seven passages in this test. Each passage is followed by several questions. After reading a passage, choose the best answer to each question and fill in the corresponding oval on your answer sheet. You may refer to the passages as often as necessary.

You are NOT permitted to use a calculator on this test.

PASSAGE I

Soil, by volume, consists on the average of 45% minerals, 25% water, 25% air, and 5% organic matter (including both living and nonliving organisms). Time and topography shape the composition of soil and cause it to develop into layers known as horizons. The soil horizons are collectively known as the soil profile. The composition of soil varies in each horizon, as do the most common minerals (see Figure 1). Figure 1 also shows the depth of each horizon and the overall density of the soil.

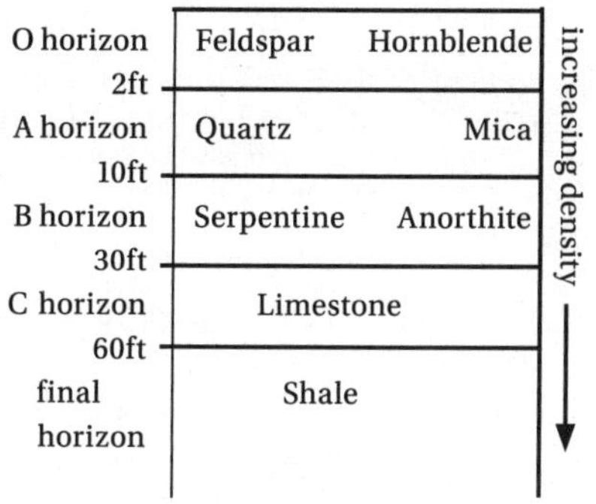

Horizon	Depth	Minerals
O horizon	2ft	Feldspar, Hornblende
A horizon	10ft	Quartz, Mica
B horizon	30ft	Serpentine, Anorthite
C horizon	60ft	Limestone
final horizon		Shale

(increasing density ↓)

Figure 1

Table 1 lists the percentages (%) of zinc and calcium in the minerals that compose soil.

Table 1		
Mineral	Zinc content (%)	Calcium content (%)
Feldspar	35–40	0–10
Hornblende	30–35	10–20
Quartz	25–30	20–30
Mica	20–25	30–40
Serpentine	15–20	40–50
Anorthite	10–15	50–60
Limestone	5–10	60–70
Shale	0–5	70–80

Table 2 shows the percentages of minerals that compose granite and sandstone, 2 rock types that are commonly found in soil.

Table 2		
Mineral	Percentage of mineral in:	
	Sandstone	Granite
Feldspar	30	54
Hornblende	2	0
Quartz	50	33
Mica	10	10
Serpentine	0	0
Anorthite	0	0
Limestone	5	0
Shale	0	0
Augite	3	3

1. An analysis of an unknown mineral found in soil revealed its zinc content to be 32% and its calcium content to be 12%. Based on the data in Table 1, geologist would most likely classify this mineral as:
   A. hornblende.
   B. anorthite.
   C. serpentine.
   D. mica.

2. Geologists digging down to the A horizon would most likely find which of the following minerals?
   F. Limestone
   G. Shale
   H. Serpentine
   J. Mica

GO ON TO THE NEXT PAGE.

3. Based on the data presented in Figure 1 and Table 1, which of the following statements best describes the relationship between the zinc content of a mineral and the depth below surface level at which it is dominant? As zinc content increases:

   A. depth increases.
   B. depth decreases.
   C. depth first increases, then decreases.
   D. depth first decreases, then increases.

4. If geologists were to drill through to the C horizon, which minerals would they most likely encounter on the way?

   F. Quartz, mica, and limestone
   G. Feldspar, shale, and serpentine
   H. Feldspar, quartz, and anorthite
   J. Hornblende, limestone, and serpentine

5. If augite is most likely found at a depth between that of the other minerals found in granite, then augite would most likely be found at a depth of:

   A. 10 feet or less.
   B. 30 feet or less.
   C. 60 feet or less.
   D. greater than 60 feet.

GO ON TO THE NEXT PAGE.

Passage II

Conductivity is the ability for a material to transmit electricity. All materials have electrical properties that divide them into three broad categories: conductors, insulators, and semiconductors. A conductor is a substance that allows an electric charge to travel from one object to another. An insulator is a substance that prevents an electric charge from traveling between objects. Substances with levels of conductivity between that of a conductor and that of an insulator are called semi-conductors. A voltmeter is an instrument used to measure voltage (see Figure 1).

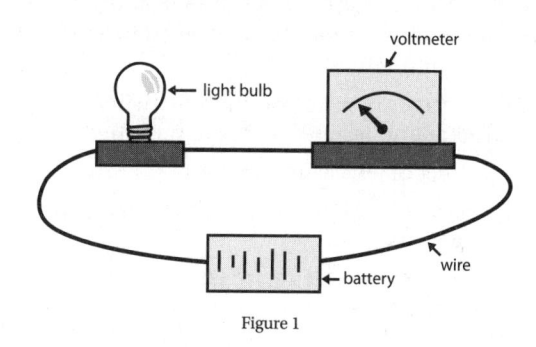

Figure 1

Three studies were executed to determine the validity of the hypothesis that a wire's conductivity increases when either the diameter or the temperature of the wire decreases.

Study 1

Wires were made from 5 different materials. Each strand of wire had a diameter of exactly 4 millimeters (mm). The strand of wire connecting the battery, light bulb, and voltmeter was 0.5 meters (m) long and was kept at a temperature of 50°C. Table 1 displays the voltage, in millivolts (mV), recorded by the voltmeter.

Table 1	
Material	Voltmeter (mV)
Silicone carbide (SiC)	4.6
Copper (Cu)	9.4
Rubber	0.0
Zinc Telluride (ZnTe)	5.2
Steel	3.5

Study 2

The conditions in Study 1 were repeated, except the diameter of the wires was decreased to 2 mm. The length between the battery and the light bulb and the light bulb and the voltmeter was held constant at 0.5 m and the wires were kept at 50°C. Table 2 displays the findings.

Table 2	
Material	Voltmeter (mV)
Silicone carbide (SiC)	6.5
Copper (Cu)	11.3
Rubber	0.0
Zinc Telluride (ZnTe)	7.1
Steel	8.6

Study 3

Study 2 was repeated at 30°C. Table 3 displays the findings.

Table 3	
Material	Voltmeter (mV)
Silicon carbide (SiC)	7.3
Copper (Cu)	12.1
Rubber	0.0
Zinc Telluride (ZnTe)	8.9
Steel	6.6

6. The scientist hypothesized that decreasing the diameter of a wire increases its conductivity. The results from the studies for each of the following materials prove the scientist's hypothesis to be true, EXCEPT the results for:

   F. silicon carbide.
   G. rubber.
   H. copper.
   J. steel.

7. According to the results of all 3 experiments, a wire made from ZnTe would have the highest conductivity with which of the following dimensions?

   A. 1 mm diameter, 0.5 m length at 20°C
   B. 4 mm diameter, 0.5 m length at 20°C
   C. 4 mm diameter, 0.5 m length at 40°C
   D. 10 mm diameter, 0.5 m length at 40°C

GO ON TO THE NEXT PAGE.

8. What would the voltmeter read if a scientist used two wires, one copper and one rubber, both with diameters of 2 mm, lengths of 0.5 m, and at 30°C, to conduct electricity to the light bulb?
   F. 0.0 millivolts
   G. 9.4 millivolts
   H. 12.1 millivolts
   J. 14.7 millivolts

9. How would the conductivity of the materials be affected if Study 3 was repeated and the temperature of the wires was increased to 100°C?
   A. The conductivity would decrease with the exception of rubber.
   B. The conductivity would remain unchanged.
   C. The conductivity would increase only.
   D. The conductivity would increase with the exception of rubber.

10. Why was the conductivity of rubber examined in all three studies?
    F. To show that rubber conducts electricity well.
    G. To determine whether the diameter and temperature of rubber affect its insulating abilities.
    H. To show that the use of rubber with any other material will increase that material's conductivity.
    J. To determine if the length of a rubber wire affects its insulating abilities.

11. Which of the following effects would be most appropriate for the scientists to test next to learn more about conductivity?
    A. The changes in wire conductivity when diameter and temperature are modified.
    B. The effect of wire color on conductivity.
    C. The effect of temperature of the wire on conductivity.
    D. The effect of different wire lengths on conductivity.

GO ON TO THE NEXT PAGE.

Passage III

From a 15-foot ladder, a scientist drops various rubber balls onto a scale that was placed on the ground to measure the force, in newtons (N), of the ball when it initially hits the scale pan (see Figure 1). A newton is defined as the unit of force required to accelerate a 1-kilogram mass at a rate of 1.0 m/sec^2. The scientist recorded the mass of the ball, its circumference, the downward force on the ball, the force on the scale from the initial hit, and the number of times the ball bounced before it reached its resting point. Table 1 displays the findings.

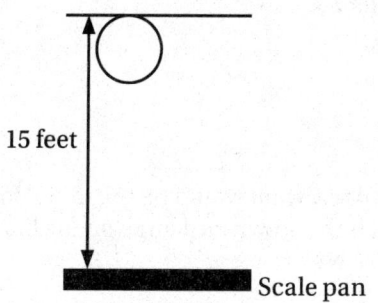

Figure 1

Table 1					
Trial	Mass of ball (g)	Circumference of ball (cm)	Downward force on ball (N)	Force on scale from initial hit (N)	# of times ball bounced before rest
1	15	10	1.0	0.79	6
2	15	10	1.5	1.19	9
3	15	10	2.0	1.58	12
4	15	10	2.5	1.98	15
5	30	10	1.0	1.58	5
6	30	10	1.5	2.38	8
7	30	10	2.0	3.16	10
8	30	10	2.5	3.96	13
9	45	10	2.0	4.74	9
10	45	20	3.0	4.29	7
11	60	10	2.0	6.32	8
12	60	20	3.0	5.71	6
13	60	30	6.0	8.63	9
14	60	40	9.0	11.47	12

GO ON TO THE NEXT PAGE.

12. How is the downward force of the ball related to the force on the scale from the initial hit in Trials 1–4?
    F. The force on the scale increases as the downward force of the ball increases.
    G. The force on the scale increases as the downward force of the ball decreases.
    H. Both the force on the scale and the downward force of the ball remain constant.
    J. There is no relationship between the force on the scale and the downward force of the ball.

13. The hypothesis that the number of times a ball bounces before reaching its resting point decreases when the mass of a ball is doubled, and the downward force on the ball is constant, is supported by which of the following trials?
    A. Trials 2 and 7
    B. Trials 3 and 4
    C. Trials 8 and 14
    D. Trials 2 and 6

14. Which of the following ranges represents the force on the pan from the initial hit of a ball with a mass of 75 g and a circumference of 40 cm that was dropped with a downward force of 9.0 N?
    F. 0.1 N to 2.42 N
    G. 2.43 N to 6.87 N
    H. 6.88 N to 11.46 N
    J. Greater than 11.47 N

15. The ball with the largest mass for its circumference is represented in which trial(s)?
    A. Trials 1–4
    B. Trial 7
    C. Trial 11
    D. Trial 14

16. Approximately what would be the initial force on the scale pan if the downward force on the ball in Trials 5–8 was 3.0 N?
    F. 2.5 N
    G. 3.3 N
    H. 4.0 N
    J. 4.7 N

GO ON TO THE NEXT PAGE.

Passage IV

Engineers designing a roadway needed to test the composition of the soil that would form the roadbed. In order to determine whether their two sampling systems (System A and System B) give sufficiently accurate soil composition measurements, they first conducted a study to compare the two systems. Soil samples were taken with varying levels of humidity (concentration of water). The concentrations of the compounds that form the majority of soil were measured. The results for the sampling systems were compared with data on file with the US Geological Survey (USGS), which compiles extremely accurate data. The engineers' and USGS' results are presented in the table below.

Table 1

Concentration (mg/L) of:	\multicolumn{5}{c}{Level of Humidity}
	10%	25%	45%	65%	80%
Nitrogen (N)					
USGS	105.2	236	598	78.1	904
System A	111.6	342	716	953	1,238
System B	196.4	408	857	1,296	1,682
Potassium Oxide (K$_2$O)					
USGS	9.4	9.1	8.9	8.7	8.2
System A	9.4	9.0	8.7	8.5	8.0
System B	9.5	9.2	9.0	8.8	8.3
Calcium (Ca)					
USGS	39.8	24.7	11.4	5.0	44.8
System A	42.5	31.4	10.4	8.0	42.9
System B	37.1	23.2	11.6	11.1	45.1
Phosphorus Oxide (P$_2$O$_3$)					
USGS	69.0	71.2	74.8	78.9	122.3
System A	67.9	69.9	72.2	76.7	123.1
System B	74.0	75.6	78.7	82.1	126.3
Zinc (Zn)					
USGS	0.41	0.52	0.64	0.74	0.70
System A	0.67	0.80	0.88	0.97	0.93
System B	0.38	0.48	0.62	0.77	0.73

Note: Each system concentration measurement is the average of 5 measurements.

---

17. At a humidity level of 25%, it could be concluded that System B least accurately measures the concentration of which of the following compounds, relative to the data on file with the USGS?
    A. Nitrogen
    B. Calcium
    C. K$_2$O
    D. P$_2$O$_3$

18. The engineers hypothesized that the concentration of potassium oxide (K$_2$O) decreases as the level of humidity increases. This hypothesis is supported by:
    F. the data from the USGS only.
    G. the System A measurements only.
    H. the data from the USGS and the System B measurements only.
    J. the data from the USGS, the System A measurements, and the System B measurements.

19. Do the results in the table support the conclusion that System B is more accurate than System A for measuring the concentration of zinc?
    A. No, because the zinc measurements from System A are consistently higher than the zinc measurements from System B.
    B. No, because the zinc measurements from System A are closer to the data provided by the USGS than the zinc measurements from System B.
    C. Yes, because the zinc measurements from System B are consistently lower than the zinc measurements from System A.
    D. Yes, because the zinc measurements from System B are closer to the data provided by the USGS than the zinc measurements from System A.

GO ON TO THE NEXT PAGE.

20. The relationship between humidity level and calcium concentration, as measured by System B, is best represented by which of the following graphs?

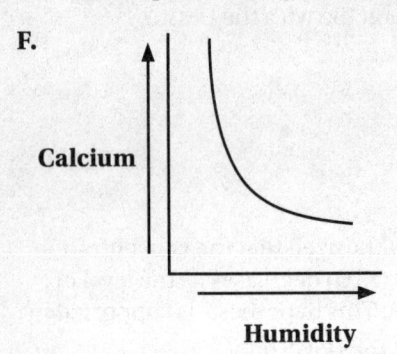

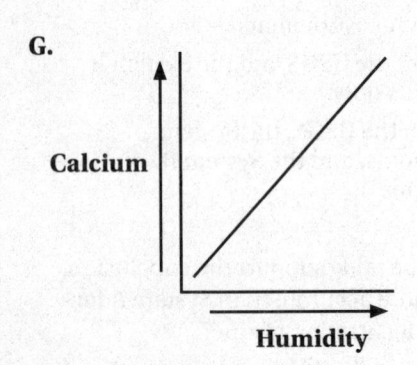

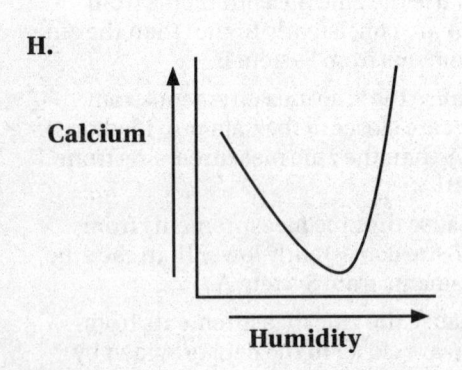

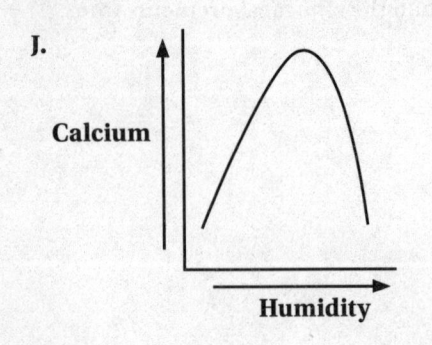

F.

Calcium vs Humidity (decreasing curve)

G.

Calcium vs Humidity (increasing line)

H.

Calcium vs Humidity (U-shaped curve)

J.

Calcium vs Humidity (inverted U-shape)

21. After conducting their comparison, the engineers used System B to test a soil sample at the future road site. They measured the concentrations, in mg/L, of selected compounds in the sample and found that they were: potassium oxide (K_2O) = 9.1; calcium(Ca) = 17.3; and zinc(Zn) = 0.57. According to the data in the table, the engineers should predict that the level of humidity is approximately:

A. 16%.
B. 37%.
C. 49%.
D. 57%.

GO ON TO THE NEXT PAGE.

Passage V

An increasing number of individuals over 50 develop type II diabetes, which occurs when the body does not produce enough insulin or when the cells ignore the insulin and as a result, the body's blood sugar level rises dangerously. Although type II diabetes occurs in people of all ages and races, it is more common in adults. Several hypotheses have been proposed to explain the cause of type II diabetes.

Dietary Hypothesis

Most Americans consume too much sugar. Sugar from food is absorbed into the bloodstream and insulin is required for the body to be able to use that sugar. In a study of individuals 18–25 years old who consumed more than the recommended amount of sugar daily, and were thus considered at risk for developing type II diabetes, it was shown that the majority had significantly elevated levels of sugar in their blood but normal levels of insulin. When these individuals received small injections of insulin once a day, their blood sugar levels decreased to more normal levels. If abundant levels of sugar are supplied by the diet, sugar-dissolving insulin injections should be given to avoid type II diabetes.

Genetic Hypothesis

Genes, which primarily come from parents and grandparents, contribute to many medical problems that individuals will experience through life. Type II diabetes mainly depends on one's genes, but also on one's lifestyle. Diabetes occurs when the pancreas produces little or no insulin at all or when the insulin it produces does not work properly. As individuals grow older, the processes of their bodies do not run as efficiently as they did when the individuals were younger. Therefore, the same behaviors may be more detrimental to a person when they are older than when they were younger, especially for those over the age of 50. This is the main reason that type II diabetes is more common in adults. Scientists compared the genetics and lifestyle of 4 groups of individuals over 50. The results are shown in the table below.

Group	Attributes	% with type II diabetes
A	One parent with type II	55%
B	Healthy lifestyle, no parents with type II	20%
C	One parent with type II and a healthy lifestyle	40%
D	Two parents with type II	70%

Exercise Hypothesis

A lack of exercise results in high body fat content, and a high body fat content does not enable the body to work efficiently. Conversely, regular weight-bearing exercise can boost the body's efficiency. One study showed that 10 weeks of weight training lowered blood sugar in adults over 50. A second study on another group of adults over 50 showed that walking 2 miles a day for 12 weeks also lowered blood sugar levels.

22. The Dietary Hypothesis would be strengthened if it were proven that high blood sugar levels are indicative of:
    F. a low efficiency of insulin.
    G. a low-sugar diet.
    H. high levels of insulin produced by the body.
    J. sugar being stored elsewhere in the body.

23. The Genetic Hypothesis best explains why type II diabetes is more common in which of the following groups?
    A. Individuals under the age of 25 as opposed to individuals over the age of 25
    B. Individuals over the age of 25 as opposed to individuals under the age of 25
    C. Individuals over the age of 50 as opposed to individuals under the age of 50
    D. Individuals under the age of 50 as opposed to individuals over the age of 50

GO ON TO THE NEXT PAGE.

24. According to the Genetic Hypothesis, adults who have had their pancreas removed should exhibit:
    F. increased blood insulin levels.
    G. decreased blood sugar levels.
    H. increased blood sugar levels.
    J. decreased body fat content.

25. Supporters of the Dietary Hypothesis might criticize the experimental results in the Exercise Hypothesis for which of the following reasons?
    A. Not enough sugar was included in the diets of the test subjects in both groups.
    B. The sugar intake of the individuals in the two groups was not monitored.
    C. The genetics of each individual in both groups should have been determined.
    D. Type II diabetes is more common in children than adults.

26. Assume that individuals with elevated blood sugar levels have a greater chance of developing type II diabetes. How would supporters of the Genetic Hypothesis explain the experimental results presented in the Dietary Hypothesis?
    F. The test subjects probably had high levels of sugar in their diets.
    G. The test subjects did not perform any weight-bearing exercise.
    H. The test subjects probably had no occurrence of type II diabetes in their genetic backgrounds.
    J. The test subjects were given too little insulin.

27. How might proponents of the Dietary Hypothesis explain the results of Group D in the Genetics Hypothesis experiment?
    A. Insulin supplements should not have been taken by this group.
    B. These individuals and both parents had an unhealthy diet that was high in sugar.
    C. More genetic background should be researched for all the groups.
    D. Not enough insulin was given to this group to affect the onset of type II diabetes.

28. The experiments cited in the Genetics Hypothesis and in the Exercise Hypothesis are similar in that each test subject:
    F. has at least one parent with type II diabetes.
    G. was given a shot of insulin.
    H. had their pancreas previously removed.
    J. was an adult over 50 years old.

GO ON TO THE NEXT PAGE.

Passage VI

Human blood is composed of approximately 45% formed elements, including blood cells, and 50% plasma. The formed elements of blood are further broken down into red blood cells, white blood cells, and platelets. The mass of a particular blood sample is determined by the ratio of formed elements to plasma, as the formed elements weigh approximately 1.10 grams per milliliter (g/mL), and plasma approximately 1.02 g/mL. This ratio varies according to an individual's diet, health, and genetic makeup.

The following experiments were performed by a phlebotomist to determine the composition and mass of blood samples from three different individuals, each of whom was required to fast overnight before the samples were taken.

Experiment 1

A 10 mL blood sample was taken from each of the three patients. The densities of the blood samples were measured using the oscillator technique, which determines fluid densities by measuring sound velocity transmission.

Experiment 2

Each 10 mL blood sample was spun for 20 minutes in a centrifuge to force the heavier formed elements to separate from the plasma. The plasma was then siphoned off and its mass recorded.

Experiment 3

The formed elements left over from Experiment 2 were analyzed using the same procedure from Experiment 2, except this time they were spun at a slower speed for 45 minutes so that the red blood cells, white blood cells, and platelets could separate out. The mass of each element was then recorded. The results of the three experiments are shown below:

Table 1

Patient	Plasma (g)	Red blood cells (g)	White blood cells (g)	Platelets (g)	Total density (g/mL)
A	4.54	2.75	1.09	1.32	1.056
B	4.54	2.70	1.08	1.35	1.054
C	4.64	2.65	1.08	1.34	1.050

29. The results of the experiments indicate that the blood sample with the lowest density is the sample with the most:
    A. plasma.
    B. red blood cells.
    C. white blood cells.
    D. platelets.

30. Why did the phlebotomist likely require each patient to fast overnight before taking blood samples?
    F. It is more difficult to withdraw blood from patients who have not fasted.
    G. Fasting causes large, temporary changes in the composition of blood.
    H. Fasting ensures that blood samples are not affected by temporary changes caused by consuming different foods.
    J. Blood from patients who have not fasted will not separate when spun in a centrifuge.

31. Which of the following best explains why the amount of plasma, red blood cells, white blood cells, and platelets do not add up to 10.5 g?
    A. Some of the red blood cells might have remained in the plasma, yielding low red blood cell measurements.
    B. Some of the platelets might not have separated from the white blood cells, yielding high white blood cell counts.
    C. The centrifuge might have failed to fully separate the plasma from the formed elements.
    D. There are likely components other than plasma, red and white blood cells, and platelets in blood.

32. From the data presented in the experiment, it is possible to determine that as total density increases, the mass of red blood cells:
    F. increases only.
    G. increases, then decreases.
    H. decreases only.
    J. decreases, then increases.

GO ON TO THE NEXT PAGE.

33. A 10 mL blood sample from a fourth individual contains 5 mL of plasma and 4 mL of formed elements. Approximately what is the mass of this blood sample?
    A. 6.5 g
    B. 9.5 g
    C. 11.5 g
    D. 15.5 g

34. The phlebotomist varied which of the following techniques from Experiment 2 to Experiment 3?
    F. The volume of blood taken from each patient
    G. The mass of blood taken from each patient
    H. The instrument used to separate the elements of the blood samples
    J. The amount of time the samples were left in the centrifuge

Passage VII

A student performed experiments to determine the relationship between the electrical conductivity of a metal rod and its length, mass density (mass per unit length of metal), and temperature.

Experiment 1

The student used several lengths of iron rods. The student weighed the rods and calculated their mass densities. The rods were then heated to the specified temperature by being held over a flame. To test the conductivity, pairs of rods were placed at opposite sides of a container containing an electrolyte solution (a solution containing positive ions with a positive electrical charge and negative ions with a negative electrical charge) and then connected to a battery. The movement of the ions in the solution was detected and displayed on the screen of an oscilloscope, where the conductivity could be measured. The results are presented in Table 1.

Table 1

Trial	Length (cm)	Mass density (g/cm)	Temperature (°C)	Conductivity ($\mu\Omega$/cm)
1	16	100	20	240
2	16	100	80	120
3	16	400	20	60
4	16	400	80	30
5	8	100	20	120
6	8	100	80	60
7	8	400	20	30

Experiment 2

The student repeated the procedure in Experiment 1, this time using rods made from silver and tungsten. The results are presented in Table 2.

Table 2

Trial	Material	Length (cm)	Mass density (g/cm)	Temperature (°C)	Conductivity ($\mu\Omega$/cm)
8	Silver	16	400	20	30
9	Silver	16	100	80	120
10	Silver	16	225	20	60
11	Silver	16	225	5	120
12	Tungsten	16	100	20	60
13	Tungsten	16	225	80	240

35. Between Trials 8 and 10, the student directly manipulated which of the following variables?
    A. The temperature of the metal rods
    B. The conductivity of the metal rods
    C. The material from which the metal rods were composed
    D. The mass density of the metal rods

36. Instead of immersing the rods in an electrolyte solution, the student could have measured conductivity by:
    F. shortening the rods to 8 cm.
    G. connecting the rods with a low-resistance wire.
    H. increasing the mass density of the rods to 200 g/cm.
    J. insulating the rods with rubber.

37. If the rods used in Trial 11 were heated to a temperature of 60°C, the conductivity would most likely be:
    A. less than 60 $\mu\Omega$/cm
    B. 80 $\mu\Omega$/cm
    C. 120 $\mu\Omega$/cm
    D. greater than 120 $\mu\Omega$/cm

GO ON TO THE NEXT PAGE.

38. Based on the results of both experiments, which of the following statements regarding the relationship between observed conductivity and the other physical variables is false?

- F. Increasing temperature increases observed conductivity.
- G. Increasing mass density decreases observed conductivity.
- H. Increasing length increases conductivity.
- J. The observed conductivity depends on the material.

39. A student is given several metal rods of identical length but unknown mass density. The rod with the lowest mass density could be found by:

- A. placing all the rods under identical temperature conditions and selecting the rod with the lowest conductivity.
- B. placing all the rods under identical temperature conditions and selecting the rod with the highest conductivity.
- C. varying the voltage until all the rods have the same conductivity and selecting the rod with the lowest temperature.
- D. selecting the rod capable of sustaining the highest temperature without melting.

40. An ordinary table lamp requires a conductivity level of less than 30 $\mu\Omega$/cm. An electrician wants to use an iron rod with the mass density 400 g/cm at a temperature of 80°C in order to conduct electricity to the lamp. Approximately what length should the rod be for the electrician to connect the table lamp to the source of the electricity?

- F. 48 cm
- G. 24 cm
- H. 16 cm
- J. Less than 16 cm

END OF TEST 2

STOP! DO NOT TURN THE PAGE UNTIL TOLD TO DO SO.